Wittgenstein on Colour

Publications of the Austrian Ludwig Wittgenstein Society New Series (N.S.)

—

Volume 21

Wittgenstein on Colour

Edited by
Frederik A. Gierlinger, Štefan Riegelnik

DE GRUYTER

ISBN 978-3-11-055480-9
e-ISBN (PDF) 978-3-11-035110-1
e-ISBN (EPUB) 978-3-11-038335-5
ISSN 2191-8449

Library of Congress Cataloging-in-Publication Data
A CIP catalog record for this book has been applied for at the Library of Congress.

Bibliographic information published by the Deutsche Nationalbibliothek
The Deutsche Nationalbibliothek lists this publication in the Deutsche Nationalbibliografie;
detailed bibliographic data are available in the Internet at http://dnb.dnb.de.

© 2017 Walter de Gruyter GmbH, Berlin/Boston
This volume is text- and page-identical with the hardback published in 2014.
Printing: CPI books GmbH, Leck
♾ Printed on acid-free paper
Printed in Germany

www.degruyter.com

Contents

List of Works of Ludwig Wittgenstein —— 2

Andrew Lugg
When and why was Remarks on Colour written – and why is it important to know? —— 1

Joachim Schulte
"We Have a Colour System as We Have a Number System" —— 21

Richard Heinrich
Green and Orange – Colour and Space in Wittgenstein —— 33

Gabriele M. Mras
'Propositions About Blue' – Wittgenstein on the Concept of Colour —— 45

Gary Kemp
Did Wittgenstein have a Theory of Colour? —— 57

Frederik A. Gierlinger
"Imagine a Tribe of Colour-Blind People" —— 67

Herbert Hrachovec
Reddish Green —— 79

Martin Kusch
Wittgenstein as a Commentator on the Psychology and Anthropology of Colour —— 93

Barry Stroud
Concepts of Colour and Limits of Understanding —— 109

Notes on the Contributors —— 119

Index —— 122

Names —— 124

List of Works of Ludwig Wittgenstein

AWL — *Wittgenstein's Lectures, Cambrdige 1932-1935, from the Notes of Alice Ambrose and Margaret MacDonald*, edited by A. Ambrose. Blackwell, Oxford, 1979.

BB — *The Blue and Brown Books.* Oxford: Blackwell, 1958; 2nd edition, 1960.

CV — *Culture and Value*, edited by G. H. von Wright, translated by P. Winch. 2nd edition, Oxford: Blackwell, 1980; Chicago: University of Chicago Press, 1980.

LFM — *Lectures on the Foundations of Mathematics*, Cambridge 1933, edited by C. Diamond. Ithaca: Cornell University Press, 1976; Chicago: University of Chicago Press, 1989.

LWL — *Wittgenstein's Lectures, Cambridge 1930-32, from the Notes of John King and Desmond Lee*, edited by Desmond Lee. Blackwell, Oxford 1980.

LWPP I — *Last Writings on the Philosophy of Psychology*, Vol. 1: Preliminary Studies for Part II of the *Philosophical Investigations*, edited by G. H. von Wright and H. Nyman, translated by C. G. Luckhardt and M. A. E. Aue. Chicago: University of Chicago Press, 1982.

LWPP II — *Last Writings on the Philosophy of Psychology*, Vol. 2: The Inner and the Outer, 1949-19S1, edited by G. H. von Wright and H. Nyman, translated by C. G. Luckhardt and M. A. E. Aue. Oxford: Blackwell, 1992.

OC — *On Certainty*, edited by G. E. M. Anscombe and G. H. von Wright, translated by G. E. M. Anscombe and D. Paul. Oxford: Basil Blackwell, 1969.

PG — *Philosophical Grammar*, edited by R. Rhees, translated by A. Kenny. Oxford: Blackwell, 1974.

PI — *Philosophical Investigations*, edited by P. M. S. Hacker and J. Schulte, translated by G. E. M. Anscombe, P. M. S. Hacker and J. Schulte. Oxford: Blackwell, 4th edition, 2009.

PR — *Philosophical Remarks*, edited by R. Rhees, translated by R. Hargreaves and R. White. Oxford: Blackwell, 1964; 2nd edition, 1975.

RC — *Remarks on Colour*, edited by G. E. M. Anscombe, translated by L. McAlister and M. Schattle. Oxford: Blackwell, 1977.

RFM — *Remarks on the Foundations of Mathematics*, edited by G. H. von Wright, R. Rhees, and G. E. M. Anscombe, translated by G. E. M. Anscombe. Oxford: Blackwell, 1956; 2nd edition, 1967; 3rd edition, 1978.

RPP I	*Remarks on the Philosophy of Psychology*, Vol. 1, edited by G. E. M. Anscombe and G. H. von Wright, translated by. G. E. M. Anscombe. Chicago: University of Chicago Press, 1980.
RPP II	*Remarks on the Philosophy of Psychology*, Vol. 2, edited by G. H. von Wright and H. Nyman, translated by C. G. Luckhardt and M. A. E. Aue. Chicago: Univer- sity of Chicago Press, 1980.
TLP	*Tractatus Logico-Philosophicus*, translated by C. K. Ogden. London: Routledge, 1922. translated by D. F. Pears and B. McGuinness. London: Routledge, 1961.
WLPP	*Wittgenstein's Lectures on the Philosophy of Psychology 1946-1947, from the Notes of P. T. Geach, K. J. Shah, and A. C. Jackson*, edited by P. T. Geach. Harvester Press, Hassocks, 1988.
WVC	*Wittgenstein and the Vienna Circle: Conversations*, recorded by F. Waismann, edited by B. F. McGuinness. Blackwell, Oxford, 1967.
Z	*Zettel*, edited by G. E. M. Anscombe and G. H. von Wright, translated by G. E. M. Anscombe. Oxford: Blackwell, 1967; 2nd edition, 1981.

Andrew Lugg
When and why was Remarks on Colour written – and why is it important to know?

> Der Philosoph will die Geographie der Begriffe beherrschen.
> (MS 137, 63a, dated 1 July 1948)[1]

The publicity material on the cover of the paperback edition of *Remarks on Colour* states that "[t]he principal theme [of the work] is the features of different colours, of different kinds of colour (metallic colour, the colour of flames, etc.) and of luminosity".[2] This description, in all likelihood approved, if not written, by the editor, G.E.M. Anscombe, is misleading. While many of Wittgenstein's remarks are devoted to the nature and motley character of colour with more than a few on the features of different colours, kinds of colour and luminosity, there is much more in the book, and readers coming to it with the publicity material in mind will almost inevitably get the wrong impression. What is said to be the principal theme is at most a minor one, and it is far from self-evident that Wittgenstein is, as also intimated on the cover, out "to destroy the traditional idea that colour is a simple and logically uniform kind of thing". An examination of when and why Wittgenstein drafted the remarks of the book shows him to be centrally motivated by questions concerning transparency, questions unmentioned in the publicity material and mostly missed or disregarded in the secondary literature.

Remarks on Colour comprises observations that Wittgenstein penned during the last fifteen or so months of his life. His literary executors – Anscombe, Rush Rhees and G.H. von Wright – apparently reckoned "the whole of this material might well be published, as it gives a clear sample of first-draft writing and subsequent selection". Anscombe allows that much of the material written not subsequently recycled is "of great interest" but decided – this is more obvious in the German version of her preface – that the chosen "method of publication involves the least editorial intervention" (Editor's Preface). Judging from how she ended up presenting the material, her thought was that Wittgenstein's remarks would be most satisfactorily published in three separate units – Part I extracted from MS 176, Part II extracted from MS 172 and Part III extracted from MS 173 (for catalogue numbers see von Wright 1993b). Not without reason Anscombe seems to have considered the relatively small number of other remarks

---
[1] The philosopher wants to master the geography of concepts.
[2] References in the text are to the published text. The numbering of the remarks is the editor's, not Wittgenstein's.

on colour in the final six items in von Wright's catalogue, MSS 172-177, incidental and safely omitted. Certainly there is nothing about colour to speak of in the remarks of MS 172, MS 173 and MS 176 not included in *Remarks on Colour* or in MS 174, MS 175 and MS 177, manuscripts mostly reproduced in *On Certainty* or volume 2 of *Last Writings on the Philosophy of Psychology*.

It would be shabby to quarrel with Anscombe's presentation of the material. Part I, the part written last, is better organised than Part III, from which it was largely drawn, and the remarks seemingly composed during the process of selection usefully fill out the discussion. (Anscombe overstates the degree to which Part I derives Part III when she describes the material in her "Editor's Preface" as a revision of earlier material "with few additions" since about a quarter of the remarks are new.) Wittgenstein's ostensible aim in Part I is to express his most recent thinking about colour simply and sharply, and it is tempting to think little is lost when the material in Part III that was not recycled is relegated to second place, Anscombe's claim in her preface about its "great interest" notwithstanding. Whether or not Part I represents Wittgenstein's considered opinion, there is plenty here for philosophers to get their teeth into, not least his treatment of the problem of which colours are primary, the nature of various kinds of impossible colour and questions regarding the phenomenon of colour blindness. Moreover placing Part II or Part III before Part I would not have made for a better book. The import of the 20 remarks of Part II is not easily fathomed, and starting with the 350 remarks of Part III would have crowded out the 88 remarks of Part I.

Still the book does not have to be read from the beginning, and there are considerable advantages to working through Wittgenstein's preliminary sketches before pouring over his last remarks on the subject. Starting with his "first-draft writing" and reserving his "subsequent selection" for later examination recommends itself, pure scholarship aside, if only because it is not at all apparent what exactly Wittgenstein is about in Part I. Focusing on the remarks that appear at the beginning of *Remarks on Colour* leaves one – if my own experience is anything to go on – with the nagging suspicion that there is more to Wittgenstein's discussion than one is seeing, that he is exercised by a problem or problems he does not, whatever his motives, explicitly state. The strategy of reading Part II and Part III first may not be the key for opening all the locks but it does, I believe, open more than a few. It provides no little insight into what lay behind Wittgenstein's renewed interest in colour, why he put pen to paper, and what he aimed to achieve. His earlier thinking illuminates his final thoughts, and it is clearer that *Remarks on Colour* is a significant document, actually a profound and challenging work of philosophy.

The question of when Wittgenstein wrote the various parts of *Remarks on Colour* is tricky. It is unclear which remarks were drafted first, in particular whether Part II is earlier or later than Part III. Only the first 130 remarks of Part III are dated, and there is no explicit indication in the text when any of the others were set down. In her "Editor's Preface" Anscombe leaves the matter hanging. She states that Part III "reproduces most of a MS book written in Oxford in the Spring of 1950" and Part I "was written in Cambridge in March 1951", but she also believes it uncertain "whether Part II ante- or post-dates Part III". These observations merit consideration since Anscombe was close to Wittgenstein at the time – he was staying in her home for all but a few months between April 1950 and February 1951, during which time the bulk of *Remarks on Colour* was doubtless written. There is, however, reason to hesitate. Apart from the fact that Anscombe sidesteps the problem of whether Part II or Part III was written first, the dates she supplies for the undated part of Part III and Part I deserve a closer look.

Anscombe seems overly cautious about the origins of Part II. As has been noted more than once, Wittgenstein is reasonably regarded as having compiled MS 172, the manuscript from which Part II derives, while at the family home in Vienna between December 1949 and March 1950, i.e. before drafting the remarks of Part III. In his catalogue von Wright states that the remarks "were probably written [...] in the early months of 1950" (Wright 1993b: 498), while Denis Paul, another scholar familiar with the manuscripts, unequivocally declares the remarks were "written in Vienna" and could not "have been written after Part III" (Paul 2007: 299). Even Anscombe herself reports in the "Preface" she wrote for *On Certainty* almost a decade before writing the "Editor's Preface" for *Remarks on Colour* that she is under the impression that Wittgenstein was in Vienna when he composed MS 172, the source of §§1-65 of *On Certainty* as well as Part II of *Remarks on Colour*. Moreover since the manuscript consists of loose sheets, it may well have been produced when Wittgenstein was in Vienna away from his manuscript books. And most telling of all, there is the fact that Wittgenstein informed Norman Malcom, von Wright and Rhees in letters penned in January 1950 that he was reading J.W. von Goethe's *Zur Farbenlehre* [*On the Theory of Colour*] (McGuinness 2008: 456-458). It is not a bad bet that this stimulated him to write the remarks of Part II, Goethe being unmistakably in the background.

The hypothesis that Wittgenstein wrote Part II in Vienna has not gone unquestioned. Thus it has been argued that the hypothesis labours under the difficulty that Wittgenstein wrote on 16th January to Malcom that he is "not writing at all because [his] thoughts never sufficiently crystallize", a confession that may be thought to cast "some doubt" on the suggestion, if not derail it altogether (McGuinness 2008: 458). This hardly settles the matter, however. In the letter Wittgenstein wrote to Rhees on 22nd January he announces that he has "written down some *weak* remarks", remarks that could well be those on colour in MS 172 reproduced as Part II of *Remarks on Colour*. This is not an implausible conjecture even granting the remarks of Part II are not noticeably weak (despite how some of them are expressed). Wittgenstein grumbled about what he was getting done when, on any reasonable measure, his work was going fine, and there is no other document in the *Nachlass* that fills the bill (or compelling reason to think the remarks in question have been lost or destroyed).[3] And why suppose Wittgenstein's comment about weak remarks in his letter to Rhees is "in slight contradiction to [...] the letter to Malcom"? His thoughts could have gelled between 16th January and 22nd January sufficiently for him to have started writing, and he could have set down some, if not all, the remarks of Part II during the week.[4]

The date of composition of Part III poses a problem since the material falls into two distinct sections. For one reason or another, Anscombe chose to publish the material drawn from MS 173 as a single unbroken unit and to leave unnoted that III, 1-130 and III, 131-350, hereafter Part IIIA and Part IIIB, were not composed at the same time (MS 173, p. 0v-31v and p. 47v-100r).[5] There is not even a dividing line in the published work between the two sets of remarks, just a cursory note in the "Editor's Preface" alerting the reader about the omission of "material on 'inner-outer' [i.e. the relationship of our inner lives to our outer behaviour], remarks about Shakespeare and some general observations about life". This is especially unfortunate since it masks that Wittgenstein broke off writing on colour to discuss psychological concepts (MS 173, p. 31v-47v; LWPP II, p. 60-71) and discourages investigation of why Wittgenstein revisited the topic subsequent to discussing psychological concepts.[6] Presumably he stopped writing on colour because he took himself to have said all he had to say about colour after composing III.130 and only later came to see that he had more to say about it. But when? While there can be little question when he began and finished writing Part IIIA, when he began and finished writing Part IIIB is harder to establish.

Part IIIA was compiled, as Anscombe takes the whole of Part III to have been compiled, in the spring of 1950, i.e. on my accounting sometime after Part II. Wittgenstein has a note at the beginning of MS 173, not reproduced in *Remarks on Colour*, to the effect that he had arrived back in England from Austria on 23rd March 1950, and it cannot seriously be doubted that he wrote the 130 remarks of Part IIIA during the next three weeks. (There are eight dates interspersed dates: "24. 3. 50" before RC III, 1; "26. 3. 50" before RC III, 4; "27. 3. 50" before RC III, 25; "28. 3. 50" before RC III, 43; "29. 3. 50" before RC III, 60; "30. 3. 50" before RC III, 70; "11. 4. 50" before RC III, 125; and "12. 4. 50" before RC III, 127.) This is a large amount of material for Wittgenstein to have written in so short a period, and since the material is not noticeably rough, it is a reasonable

---

[3] There is nothing on colour in MS 170 and just a parenthetical remark about reddish-green leaves in MS 171 (LWPP II, 59). I believe and shall argue shortly that the only remarks of consequence on colour in MS 169 were drafted after the remarks of Part II.

[4] It is also somewhat misleading to suggest that Wittgenstein's remarks on colour were "inserted in a longer set of reflections, mostly on themes to do with certainty" (McGuinness 2008: 458). The sheets reproduced in *Remarks on Colour* are separated from the sheets reproduced in *On Certainty*.

[5] "The Part III notes come in two sections" (Paul 2007: 301). For the division of Part III into Part IIIA and Part IIIB, see Rothhaupt 1996: 380.

[6] It is not impossible that some or all of Part IIIB was composed at the same time or even after MS 174 or MS 175.

conjecture that he was working from notes. It is hard to believe anyone, even someone capable of writing as fast as Wittgenstein, could pen all but nine of the remarks in just five days. Still we can be pretty sure that the remarks themselves postdate rather than antedate the remarks of Part II – and all the more so when the nature of the remarks themselves, yet to be discussed, is taken into account.

Unlike the remarks of Part IIIA, the remarks of Part IIIB are undated, and there is no sign in the text of exactly when they were drafted beyond the fact that they occur later in the manuscript and hence must have been drafted after 12th April 1950. The only clue as to how long after is that von Wright reports that colour was the main topic of conversation when Wittgenstein was staying at his home in late April and early June 1950 (Wright 1993a: 478). While hardly decisive, this suggests, if only weakly, that the remarks were set down at least in part soon after the remarks of Part IIIA. It is even possible that Wittgenstein composed some of Part IIIB before the end of April, a remark in MS 174 being dated 24th April 1950 (MS 174, p. 2r; LWPP II, p. 81). In any event the order in which Part II, Part IIIA and Part IIIB were written can be safely taken to match the order in which the manuscripts appear in the catalogue. A lower catalogue number does not necessarily signify earlier composition – when a document was written was only one consideration at play when the catalogue was drawn up – but it is practically certain that the remarks derived from MS 172 precede the remarks drawn from MS 173 and Part IIIA precedes Part IIIB.

The discussion of Part I, like the discussions of Part II and Part IIIB, is undated. It is entirely uncontroversial, however, that it was composed after Part IIIB, this being the main source of many of its remarks (there are no remarks in Part I deriving from Part II and just a few remarks from Part IIIA). Since Wittgenstein died on 24th April 1951, the material must have been put together in mid-to-late 1950 or the first months of 1951. Less obvious, however, is whether it was put together before or after 1st January 1951. It has been suggested that it was compiled some time in 1951 (Nedo 1993: 145), even as late as March 1951 (Anscombe, "Editor's Preface") or – as the Bergen Electronic Edition has it – after 1st April 1951. At first blush, a 1951 date is reasonable since the remarks in MS 176 immediately following the remarks published as Part I are dated 21st March 1951. But against this suggestion there is the awkward fact that the final remarks of MS 175 are also dated 21st March 1951, and it is beyond belief that Wittgenstein could have written in a single day the 88 remarks of Part I of *Remarks on Colour* many of which are new, along with nine fairly substantial remarks on certainty (MS 175, p. 74v-78v; OC, §§417-425). A more likely hypothesis, I venture to suggest, is that Wittgenstein began MS 176 before completing MS 175 and used the empty pages of MS 176 on running out of room in MS 175. He had gone back and forth between manuscript volumes more than once before, and he could have compiled Part I just a few months, possibly a few weeks, after writing Part IIIB rather than the best part of a year afterwards. If forced to guess when he compiled the material, I would say he compiled it before going to Norway in October-November 1950, while there, or soon after returning to England.[7]

Having considered when Wittgenstein wrote the various parts of *Remarks on Colour*, I turn to the question of what could have spurred him to discuss colour in 1950 and why he set down the remarks of Part II (and later still the remarks of Part IIIA, Part IIIB and Part I). True to form, he does not say what he is aiming to show but plunges straight in. The sole tipoff regarding his decision to re-examine the topic – besides what can be gleaned from the discussion of Part II itself – is that he reports in his January 1950 letters to Malcom, von Wright and Rhees that has been finding Goethe's *Zur Farbenlehre* worth thinking about. It has to count for something that he informs Malcom that the book, "with all its absurdities, has very interesting points", informs von Wright that the book is "partly boring and repelling but in some ways also *very* instructive and philosophically interesting", and informs Rhees that while the book "attracts and repels", "[i]t's certainly philosophically interesting". What might have caught his eye? Since he does not say in his three letters or, as far as I know, anywhere else what he finds "philosophically interesting", there is no alternative to looking for hints in the text.[8]

While it has been suggested that "[i]t would be easier to tell a consistent story about [Wittgenstein's] colour ideas without [Part II]" (Paul 2007: 299), I believe the material is singularly revealing both regarding Wittgenstein's thinking about colour and his decision to write again about it. Wittgenstein begins, not uncharacteristically, with a terminological observation and notes that "the colour-impression of a surface" can be equated with "the composite of shades of colour, which produces the impression" (RC II, 1).[9] Next Wittgenstein introduces what seems to be his leading idea, the idea he mainly wants to emphasise. He writes: "blending in white removes the *colouredness* [Farbige] from the colour; but blending in yellow does not. – Is that the basis of the proposition that there can be no clear transparent white?"(RC II, 2).[10] This in turn prompts him to ask: "What then is the essential nature of *cloudiness* [das Wesentliche des *Trüben*]?" and declare: "[R]ed or yellow transparent things are not cloudy; white

---

[7] Though the remarks on certainty in MS 176 are continuous with the remarks on colour, they seem – judging from the handwriting – to have been drafted at different times. Von Wright has MS 176 down in his catalogue as from "1950; 10 March-21 April 1951" (Wright 1993b: 489).

[8] The question of why Wittgenstein began reading Goethe's book seems unanswerable. I see no reason, however, to think he opened it "exactly with the intention of spurring himself to philosophize" (Monk 1990: 561).

[9] Since MS 172 comprises loose-leaf pages, RC II, 11-20 could have been composed before RC II, 1-10. Internal evidence, however, counts against this possibility (cf. Rothhaupt 1996: 377-379). For a contrary view, see Salles 2001.

[10] To scotch a possible misunderstanding I should point out that Wittgenstein is not saying when white is added, something colourless will sooner or later result, just noting that the more white that is added, the less coloured the colour will be (and hence will eventually become opaque white).

is cloudy" (RC II, 4). Moreover a few remarks further along he adds: "'The blending in of white obliterates the difference between light and dark, light and shadow [*Hell und Dunkel, Licht und Schatten*]'; does that define [*bestimmt*] the concepts more closely? Yes, I believe it does" (RC II, 9). What he is saying – I fancy this is the thought that gelled in January 1950 – is that there is no such colour as transparent white since white is essentially cloudy.

If RC II, 1-10 can, as I belive, be regarded as indicating how Wittgenstein was thinking in early 1950, he returned to the topic of colour because he became puzzled by the impossibility of transparent white. He is to be regarded as taking up the topic of colour between writing to Malcom on 16th January and writing to Rhees on 22nd January because he noticed something in *Zur Farbenlehre* about transparent white. I picture him thinking he needed to examine closely why white is (logically) never transparent and, more generally, why some colours can (logically) be transparent and some not, this being something he had not previously looked into. It would be rash to regard his remarks at the beginning of Part II as proving he wrote them because he came to think it incumbent on him to explain why white is invariably opaque while red, yellow and other spectral colours can be either transparent or opaque. But there seems no other reason for him to have discussed the topic and every reason to think his reading of *Zur Farbenlehre* had something to do with it. This would account for his newfound interest in colour and is, I submit, a good working assumption, at least pending evidence pointing the other way.[11]

Be this as it may, it is hard to miss that the discussion of RC II, 1-10 is in the spirit of Goethe's treatment of colour in *Zur Farbenlehre*. The suggestion that blending in white removes colouredness, the idea that cloudiness "conceals forms because it obliterates [*verwischt*] light and shadow [*Licht und Schatten*]" (RC II, 5) and the claim that white "does away with darkness [*Dunkelheit*]" (RC II, 6) are all Goethean in spirit. While Wittgenstein had little time for Goethe's Aristotelian view of colour as caused by the interaction of light and darkness at light/dark boundaries, he allies himself with Goethe – at the level of concepts – when he connects whiteness with cloudiness and speaks of blending in white as obliterating the difference between light and dark, light and shadow. For him the interconnections among the notions of "white", "cloudiness", "light-dark" and "light-shadow" that Goethe stresses go a long way to "defin[ing]" the concepts. (At RC II, 10 Wittgenstein adds that were anyone not to "find it to be this way, it wouldn't be that he had experienced the contrary, but that we wouldn't understand him"). Moreover besides offering an analysis cast in terms of concepts of the sort Goethe favoured, Wittgenstein explicitly observes that "[p]henomenological analysis (as e.g. Goethe would have it) is analysis of concepts" (RC II, 16).

---

[11] It is important to remember that Wittgenstein takes it to go without saying that there is logically, not just physically, no such colour as transparent white. Cf. my 2014 article.

The impossibility of transparent white would have struck Wittgenstein as deserving special attention, one of his major concerns, early and late, being to show that logical impossibilities are syntactical rule, linguistic, grammatical. It was a leitmotif of his philosophy that there is no possibility or impossibility that is not at root conceptual, and he would have thought – on pain of exposing his philosophical vision to serious criticism – that he needed to explain the logical impossibility of transparent white. What he requires, he would have realised, is a "grammatical rule" that accounts for its incongruity. Since this incongruity cannot be explained by the simple expedient of noting that white is an opaque colour – and doubly so since the German for "opaque" is "*undurchsichtig* [not transparent]" – he would have felt he needed an analysis of white that entails that it, unlike red, yellow, green and blue, is never transparent. For him "White surfaces are opaque" is on a par with "Circles are constructible through three non-collinear points", the one because "transparent" and "white" go hand in hand, the other because the same is true of "circle", "collinear" and "point".[12]

One reason the obervation that white is essentially opaque would have set Wittgenstein back on his heels is that it is in sharp conflict with the conception of colour grammar he had been working with for the previous two decades. Prior to 1950 he had taken the logic of colour concepts to be captured by various representational devices – the colour circle, the colour octahedron and the colour double cone, in particular. Thus in *Philosophical Remarks*, a set of remarks compiled in 1930, he writes: "[T]he colour octahedron [i.e. a double pyramid with white and black represented at the apexes and red, blue, green and yellow at the corners of the base] is grammar" (PR, 39).[13] This conception has considerable merit but falls short when it comes to transparent colours – and likewise for the other representations Wittgenstein mentions. Since such representations make no provision for transparent colours, only for (opaque) surface colours, they do not explain the difference between transparent white and transparent red. In 1950 Wittgenstein would have recognised that however well the colour octahedron and the like capture the relationships among spectral colours, they require supple-

---

[12] In a book heavily influenced by Wittgenstein's thought, W. H. Watson writes: "What the proposition ['A circle can be drawn through any three points, which are not collinear'] asserts is a rule of logical grammar about the words 'circle' and 'point' " (Watson 1938: 11).

[13] Wittgenstein also refers to the colour octahedron as "a grammatical representation, not a psychological one" (PR, 1) and adds it "is grammar, since it says that you can speak of a reddish blue but not of a reddish green, etc" (PR, 39). (This is because red and blue are represented as adjacent, red and green as opposed.) Also note that Wittgenstein reportedly asserted in a lecture in February 1930 that both the colour octahedron and Euclidean geometry are "a part of grammar" (LWL, p. 8).

mentation or replacement.¹⁴ In particular he could no longer accept that the colour octahedron provides an adequate "bird's-eye view [*übersichtliche Darstellung*] of the grammatical rules [governing colour]" (PR, 1).¹⁵

Though Wittgenstein does not expressly point out that the colour octahedron fails to capture the grammar of colour in its entirety, he remains wedded to the general philosophical conception of colour that undergirds his remarks about the colour octahedron in *Philosophical Remarks*. RC II, 2-10 provide an analysis of whiteness that explains why it is never transparent or – what comes to the same thing – why it is definitive of white that it is opaque. As in *Philosophical Remarks* and other earlier work, he hopes to demonstrate that a necessity is linguistic, not empirical (or metaphysical), and continues to hold that an "analysis of concepts […] can neither agree with nor contradict physics" (RC II, 16). To his way of thinking the proposition about blending in white mentioned in RC II, 2 cannot be "a proposition of physics" and nobody should "believe in a phenomenology, something midway between science and logic" (RC II, 3).¹⁶ While he no longer takes colour concepts to be as compactly representable as he had taken them to be in *Philosophical Remarks*, he still thinks their grammar is representable and believes a satisfactory (grammatical) representation would state what can and cannot be sensibly said about colour. He does not disown the idea of the colour octahedron as encapsulating the grammar of surface colours, just insists that accommodation be made for transparency and transparent colours.

At this juncture I can imagine it being objected that I am wrong about the development of Wittgenstein's thinking about colour since the impossibility of transparent white is discussed in MS 169, a manuscript that appears in the catalogue before MS 172 (MS 172, 77v-80v; LWPP II, p. 47-48). If MS 169 was compiled in the first half of 1949 (Wright 1993b: 488) or in the summer of the same year or soon after (Gennip 2003: 131), the discussion of transparent white in Part II of *Remarks on Colour* was not Wittgenstein's first discussion of the subject but one penned months after he had initially treated it. Before jumping to conclusions, however, it should be noted that MS 169 may not have been produced all at once and the remarks on transparent white, which occur in the final pages of the manuscript, may have been written after the remarks in

---

14 It is no objection that Wittgenstein says in *Philosophical Remarks* that the colour octahedron provides a "rough representation" (PR, 1). His thought is that this representation charts the main contours of our use of colour words in much the same way that elementary logic charts the main contours of our use of "and" and "not". He was not budgeting for transparency and would not have regarded "transparent" as comparable to "not not" used to indicate strong disagreement.
15 In my view Lee rightly stresses that transparent white is a major topic of concern in *Remarks on Colour* but misses that it poses a major problem for the conception of the colour octahedron as grammar (cf. Lee 1999: 231) while McGinn rightly notices that the colour octahedron over-idealises but misses that it does not explain the impossibility of transparent white (cf. McGinn 1991: 442).
16 Here, evidently, "phenomenology" is to be understood different from what how it is understood in RC II, 16. Wittgenstein is not equating it with the analysis of concepts.

Part II, indeed after the remarks of Part IIIA. They differ in tone and substance from the few brief remarks on colour earlier in the manuscript, and it is hardly impossible that Wittgenstein was using spare pages to jot down a thought or two.[17] In addition, as will soon become evident, the discussion of transparent white in MS 169 is closely allied with the discussion in Part IIIB.

But am I not stretching it when I suggest that Wittgenstein was moved to write again on colour by what he read in *Zur Farbenlehre* about transparent white? There is little to be concluded from the fact that Wittgenstein speaks of himself as reading Goethe's book and in *Remarks on Colour* refers more frequently to Goethe than to any other thinker. He does not explicitly mention Goethe's views about transparency, only recommends reading him as analysing concepts and expressing conceptual truths (compare RC III, 125 and RC I, 70-71). Nor does Wittgenstein cite, even allude to, Goethe's characterisation of white as "the simplest, brightest, first, opaque occupation of space" (Goethe 1970, #147), his claim that "[t]ransparency itself, empirically considered, is already the first degree of the opposite state" (Goethe 1970, #148) or his thesis that it is a "tendency of a transparent medium to become only half-transparent" (Goethe 1970, #238). The fact that he says in his letter to Rhees that he has been reading "parts of Goethes [sic] Farbenlehre" is no guarantee that he had been reading the parts touching on transparency and whiteness.

It would be premature, however, to discount the hypothesis that Wittgenstein was motivated to discuss transparency and transparent white by reading *Zur Farbenlehre*. However shaky the suggestion that he read Goethe's remarks on the topic, he had to have read the letter from Philipp Otto Runge that Goethe reproduced as an appendix to his book, a letter in which transparency and transparent white figure prominently. Runge is the most cited writer after Goethe in *Remarks on Colour* and his letter is quoted, albeit in Part IIIA and Part I, not in Part II. Thus Wittgenstein writes: "Runge to Goethe: 'If we were to think of a bluish orange, a reddish green or a yellowish violet, we would have the same feeling as in the case of a southwesterly northwind'. Also: what amounts to the same thing, 'Both white and black are opaque or solid. [...] White water which is pure is as inconceivable as clear milk'" (RC III, 94, ellipsis in the original; also at RC I, 21, slightly modified, as from "Runge"). Given that Wittgenstein took "reddish green" to be linguistically anomalous, he could not but have been struck by Runge's comparison of "reddish green" with "a southwesterly northwind" and "transparent white" with "reddish green".[18]

---

[17] There are lines in MS 169 before and after the material on transparent white, in fact unusually many dividing lines at the end of the manuscript. On the manuscript itself cf. Rothhaupt 1996: 369-372, and note that von Wright and Nyman suggest in their "Editors' Preface" to *Last Writings* that the manuscript falls into two parts, namely 2-41 and 41-49. (LWPP II, p. ix)

[18] It is worth noticing that the compass perspicuously represents direction and treats north and southwest as opposed in much the same way that the colour octahedron perspicuously represents colour and treats red and green as opposed.

Accepting that Wittgenstein wrote RC II, 1-10 of *Remarks on Colour* with the object of explaining the impossibility of transparent white, the question of why he wrote RC II, 11-20 and Part IIIA is easily answered. Having explained to his own satisfaction why white is essentially opaque, he would, naturally enough, have taken it upon himself to re-examine the problem of colour as if for the first time and without preconception. He had fallen into the trap of touting an oversimplified conception of colour once and would not have wanted to fall into the same trap again. This would have been reason enough for him to proceed in the second half of Part II to discuss various topics on more or less loosely related to transparent white and to broaden the focus in Part IIIA to encompass additional sorts of colour. It made sense for him to consider, as he does in Part II, how a painter would depict objects through coloured glass, how things look in different sorts of light and whether "white light" is an intelligible concept, and discuss, as he does in Part IIIA, the use of the "-ish"-suffix, the notion of a pure colour and the concepts of brown and luminous grey.[19]

So much for Part II and Part IIIA of *Remarks on Colour* and the role of the discovery of transparent white in Wittgenstein's thinking in early 1950. The question that now arises is why, after penning a fair number of remarks on "the inner and outer", he suspended writing on this and and began again to write on colour, i.e. why he wrote the remarks of the rest of MS 173, the material comprising Part IIIB. One possibility is that he had set down all he had to say about psychological concepts and decided to continue exploring colour concepts for want of something better to explore. Alternatively he may have decided to excerpt and reorganise what he had written in Part II and Part IIIA. And it is possible too that he came to think he had something extra or different to say about colour, even perhaps that he needed to correct what he had said in Part II or Part IIIA. Unfortunately, yet again, he does not explain himself. Prior to examining Part IIIB itself, all that can be said for sure is that he was moved to think again about colour, to put pen to paper and discuss it rather than go on studying psychological concepts or return to the investigation of certainty he had begun in MS 172.

Turning to the text itself, it is clear from the first couple of pages of Part IIIB that Wittgenstein is neither developing thoughts he had had while writing on "the inner and outer" nor embellishing, reworking, bringing together or reordering what he had written in Part IIIA. He does not take up the topic of colour where he left it nor does

---

19 The topic of transparency is not neglected in Part IIIA, just treated more cursorily. In addition to quoting Runge at RC III, 94, Wittgenstein couples transparency with saturation (RC III, 14), reconsiders the task of painting transparency (RC III, 23), links transparency with black and white (RC III, 24), contrasts transparency with cloudiness (RC III, 70) and notes that Runge observes that "there are transparent and opaque colours" (RC III, 76). Still I would question whether Wittgenstein's "discovery" of the opacity of white structures the whole document and informed his remarks from beginning to end (cf. Lee 1999: 217). Rather it seems to have prompted him to discuss colour again. Also compare Waismann: "[Wittgenstein] has the marvellous gift of always seeing everything as if for the first time." (Waismann 1979:26)

he recycle remarks, his usual practice when separating wheat from chaff.[20] The remarks of Part IIIB are new, and Wittgenstein is most charitably read as returning to the drawing board because he had fresh thoughts to express.[21] Sometime after completing Part IIIA, he seems to have come round to thinking that he had not properly understood the concept of transparency and that his account of the impossibility of transparent white leaves something to be desired. In Part IIIB he discusses the nature of transparency itself, something he had not explicitly done in Part II or Part IIIA, and he supplements his explanation of the phenomenon with a markedly different explanation of the impossibility of transparent white from the one he had floated earlier. Roughly stated, his remarks on transparent white in Part IIIB differ from the remarks in Part II and Part III in that they focus on transparency rather than on whiteness, and the influence of Goethe is much less in evidence.

Wittgenstein seems to have come round to believing – this would be sometime after 12th April 1950 – that he had misconstrued the notion of transparency, at least placed the emphasis in the wrong place. He does not deny that the opacity of white is connected with cloudiness but no longer sees the absence of cloudiness as the defining characteristic of transparency. Perhaps because he had been overly influenced by Goethe's discussion of colour in *Zur Farbenlehre* and had accorded too much weight to the thought that "[b]lack and white themselves have a hand in the business, where we have the transparency of a colour" (RC III, 24), he had treated the opacity of opaque surfaces instead of the transparency of transparent ones as crucial. As he now sees it, transparency is a matter of "see-throughness", not the absence of cloudiness, and a transparent white glass is impossible because of the essential "non-see-throughness" of white rather than to its essential cloudiness. (Wittgenstein's missing that that the German for "transparent", "*durchsichtig*", literally means "through-viewable" is not especially odd, English speakers being just as liable to overlook the etymology of "transparent" [literally "appearing across"].)

There is in any case more than a few remarks in Part IIIB underlining the connection between transparency and "depth" and "behindness". This is something Wittgenstein seems not to have appreciated earlier and he now underlines the point. He identifies transparency with seeing "something as lying *behind* the glass" (RC III, 141), says "[t]he various 'colours' do not all have the same connexion with three-dimensional vision" (RC III, 142), observes that this has to do with "the connection between three-dimensionality [and] light and shadow" (RC III, 144), notes that "'[t]ransparent' could be compared with 'reflecting'" (RC III, 148) and states that "[t]ransparency and reflection exist only in the dimension of depth of a visual image" (RC III, 150). Similarly

---

[20] There is very little overlap between Part IIIB and Part IIIA. The "correspondences" listed in Rothhaupt (1996: 426) are at most rough correspondences of theme.
[21] Paul takes Wittgenstein to make a "new start" (Paul 2007: 301) but does not explain what this consists in or why Wittgenstein started afresh.

a bit further along in Part IIIB he says straight-out: "The impression of a coloured transparent medium is that something is behind the medium. Thus if we have a thoroughly monochromatic visual image, it cannot be one of transparency" (RC III, 172). And later still he adds: "The colour of a transparent glass could be said to be how a white light source, seen through it, would appear" (RC III, 183, translation revised slightly). In these remarks the difference between transparent and opaque colours is traced to their connection with "depth".

The conception of transparency as involving "three-dimensionality" augurs a major shift in Wittgenstein's thinking about colour concepts. Pre-1950 he had treated the language of colour as an autonomous department of language, one altogether separate from the language of spatial position (and the concepts of up-down, in front-behind, right-left, etc.). Throughout the 1930s and 1940s he adhered to the view adumbrated in *Philosophical Remarks*, namely: "It is clear that there isn't a relation of 'being situated' which would hold between a colour and a position, in which it 'was situated'. There is no intermediary between colour and space. Colour and space saturate one another" (PR, p. 257; also compare TLP, 2.0131 and 2.0251). Only in 1950, when he came to reflect on the impossibility of transparent white, I am speculating, did he notice that the logic of colour concepts is intimately related to the logic of spatial concepts in the form of the "dimension of depth". This was no small shift in viewpoint. The colour octahedron had been his prime example of a perspicuous representation, and colour language his stock example of an autonomous subdivision of our language.[22]

Wittgenstein would have realised that regarding transparency as essentially connected with "behindness" meant he needed a new account of the impossibility of transparent white. Noting that white is cloudy and cloudy surfaces function as opaque barriers leaves out the "dimension of depth", and in Part IIIB Wittgenstein takes up the task of finding a substitute explanation. He considers why it is senseless to speak of seeing something as lying behind a white surface and why the concept of white is different from the concept of red in point of its "see-throughness". Why is it, he in effect asks, that "behindness" is always absent in the case of white, only sometimes absent in the case of red, a question more easily raised than answered? Whence in Part IIIB he comes back to the problem, actually (though rarely noticed) comes back to it several times and has several stabs at explaining the difference. While he touches on other topics with the object, as I understand him, of further exploring the ins-and-outs of colour language, much of his discussion, especially in the first hundred or so remarks

---

22 Arguably, the analysis of opacity in terms of cloudiness provided in Part II involves no more reference to spatial or similar concepts than the concept of "darkness" and no more signals a radical departure from the views of 1930-1949 than an analysis of brown as "a 'reddish-blackish-yellow' " (RC III, 126). McGinn observes that Wittgenstein believed we "overestimat[e] [...] the degree of independence of colour concepts and spatial concepts" (McGinn 1991: 442) but makes nothing of the point.

of Part IIIB, is devoted to accounting for the impossibility of transparent white given his revised conception of transparency.[23]

The discussion of Part IIIB begins almost immediately with remarks on the topic of transparent white. After some brief preliminary observations (RC III, 131-135), Wittgenstein argues that there can be no transparent white glass since black and white seen through such a white glass, were one possible, would appear the same, not as they should appear through a transparent glass (RC III, 136). He writes: "By analogy with other colours, a black drawing on a white background seen through a transparent *white* glass would have to appear unchanged as a black drawing on a white background. For the black must remain black and the white, because it is also the colour of the transparent body, remains unchanged". Otherwise put, a transparent white glass is impossible since the opposite assumption reduces to absurdity (compare proving there can be no greatest prime number by showing that were there a greatest prime there would be an even greater one). The key point is that transparent white is ruled out by virtue of logic, not by virtue of how the world happens to be. As in Part II, Wittgenstein appeals to what he takes to be a rule of grammar, his thought being that, given how we think and speak of transparency, white surfaces are never transparent.

Wittgenstein develops much the same argument in a later remark. At RC III, 173 he writes: "Something white behind a coloured transparent medium appears in the colour of the medium, something black appears black. According to this rule a black drawing on white paper behind a white transparent medium must appear as though it were behind a colourless medium". As before, he argues by *reductio ad absurdum* that transparent white is impossible since a transparent white surface would appear colourless, not both transparent and white. The argument is again logical, the conclusion being understood to follow from the "rule" that through a coloured medium white takes on the colour of the medium and black stays black. In fact Wittgenstein points out that this last observation (there is, as the editor notes, an arrow in the manuscript pointing to it) is "not a proposition of physics, but rather a rule of the spatial interpretation of our visual experience" or, what amounts to the same thing, "a rule for painters" to the effect that white objects have to be painted the colour of a surface for the surface to appear transparent.

In subsequent remarks Wittgenstein attacks the problem from a slightly different angle. He first considers how objects appear through transparent green, red and other coloured glass (compare RC III, 175, 179 and 184), then argues: "If a pane of green glass gives the things behind it a green colour, it turns white to green, red to black, yellow to greenish yellow, blue to greenish blue. The white [transparent] pane should, therefore, make everything whitish, i.e. it should make everything *pale*; and, then why shouldn't it turn black to grey? – Even a yellow glass makes things darker, should a

---

[23] In what follows I outline Wittgenstein's rather complicated exploration of transparency and transparent white in Part IIIB and Part I. For more detailed discussion see my 2014.

white glass make things darker too?" (RC III, 191). This line of argument – that transparent white glass should make white things appear both lighter and darker, an out-and-out impossibility – is one Wittgenstein seems to have found especially compelling (the argument or a close variant of it is developed in RC III, 192-194 and 243). Moreover he acknowledges that he is appealing to a rule of grammar. He writes: "White seen through a coloured glass appears with the colour of the glass. That is a rule of the appearance of transparency" (RC III, 200). Once more the discussion is logical, Wittgenstein's thought being that it follows – given the logic of colour concepts – that white is essentially opaque.

Much the same argument is canvassed in MS 169, a compelling reason, surely, for thinking that the remarks on transparent white in this supposedly earlier manuscript were written around the same time as the remarks in Part IIIB and after the remarks in Part II and Part IIIA. Expressing the point, if anything, more sharply, Wittgenstein writes: "Flat black seen through yellow glass is black, white is yellow. Therefore analogously black must appear black seen through transparent white, and white white, i.e. just as through a colourless glass. – Is red now to appear whitish? i.e. pink? But what will a dark red, which tends towards black, appear as? It should become a blackish pink, i.e. a greyish red, but then black probably will not remain black" (LWPP II, p. 47). Moreover he avers: "White seen through yellow wouldn't become yellowish-white but yellow. And yellow seen through white – should it become whitish-yellow or white? In the first case the 'white' glass acts like colourless glass, in the second like opaque glass" (LWPP II, p. 48). In other words, when one thinks through how the spectral colours would appear behind white transparent glass, mindful of how white appears through yellow or red transparent glass, one ends up stymied.

Returning to the remarks of Part IIIB, the question arises – assuming Wittgenstein began writing them with the object of conveying new thoughts about transparency and transparent white – of why he discusses the matter at such length rather than simply states how he now sees things. The answer cannot, I think, be simply that he was enamoured by his treatment of the problem and could not resist repeating himself. This would be out of character, and there is much in Part IIIB that can hardly be counted as mere repetition. Moreover there seems to be a more compelling reason. He is, I believe, most plausibly read as revisiting the topic because he was not fully satisfied that he was right about how a white transparent glass, were one possible, would behave. At RC III, 137 he entertains the possibility of "a glass through which black looked like black, white like white, and all the other colours appeared as shades of grey; so that seen through it everything appears as though in a photograph". At RC III, 175 he asks: "[W]hy shouldn't we want to call [a glass through which everything appeared in shades ranging from white to black] *white*? Is there anything to be said against doing this; does the analogy with glass of other colours break down at any point?" And at RC III, 185-186 he wonders whether a "white" pane should, like a green pane, give "things its colour" and whether – accepting that a "thin layer of a coloured medium colours things only weakly" – we should suppose that "a thin 'white' glass

[…] doesn't quite remove all their colour". Nor is this the end of the matter. Wittgenstein continues to hesitate (see, e.g., RC III, 205, 208 and 242).

As for Part I, it needs noting straight off that the remarks on the topic of transparency and transparent white, like the remarks on other topics, are mostly extracted from Part IIIB. There is nothing on transparency deriving from Part II or from Part IIIA except the quotation from Runge at RC I, 21.[24] Wittgenstein mostly recycles remarks about transparency and transparent white verbatim, his apparent aim being to reorganise and pare down what he had earlier written. While he devotes more remarks to transparency than to any other topic, some 22 of the 88 remarks of Part I being, on a conservative count, on this subject, he does not clarify the matter to any significant degree but leaves it practically in the same state as in Part IIIB. He repeats what he said there about the nature of transparency, restates his explanation of transparent white and expresses the same hesitations about how objects should appear through a transparent white glass. It is not true, as has been suggested (cf. Brenner 1999: 117-127) that Wittgenstein comes to a definite conclusion. Rather he confines himself to restating thoughts he seems to have believed worth preserving and questions he believed worth pursuing.

It remains to consider the importance of knowing when *Remarks on Colour* was written and why. There are four lessons I draw from the present discussion. The most obvious is that a study of Wittgenstein's remarks in the order of composition illuminates the nature of his investigations of colour and puts the reader in a better position to appreciate the discussion of Part I. It brings out the central role of transparency and transparent white in *Remarks on Colour* and exposes the error of taking him to be primarily concerned with the impossibility of reddish-green, primary colours, the relationship of lightness and darkness or other topic that he had previously discussed. In particular reading Part II first reduces the chance of his treatment of transparent white being regarded as of secondary importance, the first appearance of the topic in Part I being at RC I, 17, a fifth of the way into the material (and after the remarks at RC I, 9-14 on reddish green). Furthermore when the material is read as written, it is difficult to overlook that Part I summarises Wittgenstein's chief results, both positive and negative, indeed provides what for all the world looks like an interim report and agenda for future inquiry.

Reading *Remarks on Colour* in the order composed rather than the order published also makes clear that Wittgenstein is not out to convey a view of the logic of colour concepts he had already adopted. When the material is approached naturally, starting at the beginning of the book, it is all too tempting to ask what he is trying to get over and to imagine he ends up with definite conclusions. Special effort is required to resist reading him as saying things he does not explicitly say, a trap less easy to fall into when

---

24 Wittgenstein could have written the sentence of RC III, 76 repeated at RC I, 17 from memory. This sentence aside, the two remarks are very different.

Part II and Part III are read first. It is not only the careless reader who is apt to interpret him as solving problems concerning colour rather than as exploring "the geography of concepts". And when Part I is regarded as free-standing, the discussion can seem exceedingly weak and the explanation of the impossibility of transparent white seem to fall short for reasons of the very sort noted by Wittgenstein himself. There is no guarantee that this will not occur when the work is read as I have intimated it is best read. But so read, it is harder to miss that Wittgenstein is grappling with the question of the grammar of transparency rather than providing an alternative grammar to the grammar he took the colour octahedron to provide, something he does not – and does not claim to – do.

A related point is that in *Remarks on Colour* Wittgenstein does not offer a philosophical theory about transparency, never mind one about colour in general. While his investigations are often described as exploratory rather than explanatory, critical rather than speculative, this crucial insight tends to be honoured more in the breach than the observance, and it helps restore the balance to read Part II and Part III before Part I. Studying Wittgenstein's words with an eye on what he says about transparent white is a useful antidote to the common practice of interpreting him as defending, unwittingly if not wittingly, substantive philosophical views. One sees he is engaged in a project of the sort he describes in the preface of the *Investigations*, i.e. as supplying "as it were, a number of sketches of landscapes which were made in the course of [...] long and involved journeyings" (PI, Preface). In fact this description applies better to *Remarks on Colour* than to the more finished works like the *Investigations*. In *Remarks on Colour* Wittgenstein is unsure of where he ought to end up, and it is clearer why he takes philosophical problems to be "of the form: 'I don't know my way about' " (PI, 123). He continues to hope to provide a perspicuous representation of colour grammar but does not know what form it will take, even whether there is a perspicuous representation to be had.[25]

And lastly I would underline that an examination of Wittgenstein's remarks in the order he wrote them belies the widespread opinion that in the mid-1930s he stopped viewing language in calculus-like terms and started viewing it in terms of language-games. The work begins with a reference to two language-games (RC I, 1; also RC III, 131), and one is easily misled into interpreting Wittgenstein as concerned with the use of colour language rather than with the logical relations among colour concepts. Even here, though, he does not endorse the so-called "language-game model" of language, the burden of his remark being that "non-temporal" propositions are categorically different from "temporal" propositions. He is orienting the discussion to come

---

25 Nor should it be forgotten that Wittgenstein attached great importance to the identification and invention of philosophical problems and, far from dismissing philosophy, he treated it with the utmost seriousness. As he is reported as having put it: "One must not in philosophy attempt to short-circuit the problems" (AWL, p. 109; dated 1934/1935).

by reminding the reader that propositions may be logical, grammatical, conceptual as well as empirical. In fact there are very few references to language-games in *Remarks on Colour* and none to speak of in the material on transparency. For the most part the discussion is devoted to the (logical) nature of the phenomenon and why transparent white is grammatically (logically) aberrant. Rather than treat language anthropologically, Wittgenstein aims to clarify "the logic of colour concepts" (RC I, 22; also RC III, 188).[26] It is by no means fortuitous that he speaks of "a sort of mathematics of colour" (RC III, 3) and takes colour to have a "geometry" (RC III, 86 and RC III, 154; also RC I, 66).[27]

I trust I shall not be read as suggesting that *Remarks on Colour* is bound to be misunderstood when read starting with Part I. Wittgenstein's references to Goethe, his different views regarding transparency and transparent white, and his hesitations are there for all to see – and it would be a decisive blow to my account of the origins of the book were they not. My point is that Wittgenstein's line of thought is far from evident when Part I is read first and Part III treated as a single continuous discussion instead of as two separate sets of remarks. Readers who treat Part I exclusively or regard the rest of the book as merely subsidiary are unlikely to see the importance of transparency in Wittgenstein's thinking and his shifting views about it. And still worse, Part II may strike them as "dull stuff" (Monk 1990: 564) and Part III reckoned as "a repetitive and rather laboured attempt to clarify the 'logic of colour concepts'" (Monk 1990: 566). Following in Wittgenstein's footsteps increases the chance of the development of his thought being properly appreciated, to say nothing of how he conceives the philosophical enterprise. Above all, it is not so easy to neglect his interest in discovering new philosophical problems or to discount the value he accorded to conceptual analysis.[28]

References

Brenner, W. H.: *Wittgenstein's Philosophical Investigations*. SUNY Press, 1999.
Gennip, K. v.: Connections and Divisions in *On Certainty*. In: *Knowledge and Belief. Papers of the 26th International Wittgenstein Symposium*, p. 129–131, Austrian Ludwig Wittgenstein Society, 2003.
Goethe, J. W. v.: *Theory of Colour*. MIT Press, 1970.

---

[26] It is, I believe, wrong to take Wittgenstein to be aiming in *Remarks on Colour* to get clear about "two distinct but related language-games", one "for describing the colours of the natural world", the other underpinned by "the precise system of colours that is defined by monochromatic samples of colour arranged on the colour wheel" (McGinn 1991: 442).
[27] What went by the board in the mid-1930s, I would argue, was not the "analysis of concepts" as such, only over-strict conceptions of conceptual analysis. For details see my 2013.
[28] I am indebted to Nuno Venturinha, Frederik Gierlinger, Mauro Engelmann and Paul Forster for comments. Thanks also to Lynne Cohen for help with the final version of the paper.

Lee, A.: Wittgenstein's *Remarks on Colour*. In: *Philosophical Investigations*, 22, p. 215–239, 1999.
Lugg, A.: Wittgenstein on Reddish Green: Logic and Experience. In: Marques, A. & Venturinha, N. (Ed.), *Wittgenstein on Forms of Life and the Nature of Experience*, p. 155–181, Peter Lang, 2010.
Lugg, A.: Wittgenstein in Mid-1930s: Calculi and Language-games. In: Venturinha, N. (Ed.), *The Textual Genesis of Wittgenstein's Philosophical Investigations*, p. 135–154, Routledge, 2013.
Lugg, A.: Wittgenstein on Transparent White. In: *Wittgenstein-Studien*, 5, p. 207–226, 2014.
McGinn, M.: Wittgenstein's *Remarks on Colour*. In: *Philosophy*, 66, p. 435–453, 1991.
McGuinness, B.: *Wittgenstein in Cambridge. Letters and Documents 1911-1951*. Blackwell, 2008.
Monk, R.: *Ludwig Wittgenstein*. Cape, 1990.
Nedo, M.: *Wiener Ausgabe. Einführung*. Springer, 1993.
Paul, D.: *Wittgenstein's Progress*. The Wittgenstein Archives, 2007.
Rothhaupt, J. G. F.: *Farbthemen in Wittgensteins Gesamtnachlaß*. Belz, 1996.
Salles, J.-C.: On *Remarks on Colour*. In: Haller, R. & Puhl, K. (Ed.), *Papers of the 24th International Wittgenstein Symposium*, p. 129–131, Austrian Ludwig Wittgenstein Society, 2001.
Watson, W. H.: *On Understanding Physics*. Cambridge University Press, 1938.
Wright, G. H. v.: Letters from Ludwig Wittgenstein to Georg Henrik von Wright. In: Klagge, J. & Nordmann, A. (Ed.), *Philosophical Occasions*, p. 459–479, Hackett, 1993a.
Wright, G. H. v.: The Wittgenstein Papers. In: Klagge, J. & Nordmann, A. (Ed.), *Philosophical Occasions*, p. 480–510, Hackett, 1993b.

Joachim Schulte
"We Have a Colour System as We Have a Number System"

> Nothing is more difficult than confronting concepts without prejudice. – For a prejudice is a system, and hence a form of understanding, though not the right one. However, to be unprejudiced means that there is nowhere for you to put down the weight you are carrying; you keep it in balance.
>
> (MS 136, p. 18)[1]

1 Introduction

The central passage I wish to discuss in this paper is quoted in my title. On 5 February 1948, Wittgenstein begins a longish series of entries in his notebook by saying that "We have a colour system as we have a number system".[2] He goes on to wonder whether these systems lie in *our* nature, or in the nature of things. A tentative reply to this question is that, surely, these systems do not reside in the nature of numbers or colours, respectively.

No reason is given for this response. Nor are we told whether in the original question the notion of a "nature of things" was intended to cover the same field as, or a wider field than, the notion of a nature of colour and number. If it was meant to be more comprehensive than these, then that would leave the range of possible answers wide open; if not, we would either have to go on to claim that the systems do lie in *our* nature or to reject the question.

Perhaps it would seem obvious to remark that whatever the nature of colour and number may be, it cannot be given entirely independently of their systems – whatever *they* may turn out to be. For if it is correct to speak of "systems" in connection with colour and number, we may regard it as absolutely reasonable to suppose that these systems are, as it were, the space in which to unfold and explain the nature of colour and number. So it is hard to see how their nature could be understood without paying

---
[1] 136:18b (= MS 136, p. 18b). Manuscripts are specified in terms of the usual von Wright numbering. Translations from manuscripts are my own.
[2] *Remarks on the Philosophy of Psychology* (= RPP) II, 426. (Volume II of RPP: ed. by G. H. von Wright and Heikki Nyman, trans. by C. G. Luckhardt and M. A. E. Aue [Oxford: Blackwell, 1980].)

attention to the systems. In other words, we would not be able to identify the nature of colour and number independently of their systems, and hence not in a position to appeal to such a supposedly independent nature for the purpose of supplying a basis for these systems.

Now, it may be that an argument along these lines captures part of what Wittgenstein has in mind. But it is an highly abstract argument, and as such not particularly illuminating, especially since we are dealing with a subject matter that seems to cry out for examples and illustrations. At any rate, Wittgenstein himself does not spell out an argument of the abstract kind mentioned. But neither does he proceed to give us much help by pointing out instructive examples. Instead, he avails himself of highly metaphorical and emphatic expressions. But while these metaphors and emphatic expressions may *suggest* something helpful, they do not achieve much by way of straightforward clarification. Thus, Wittgenstein's example of the alleged impossibility of forming an image of bluish yellow is commented on by saying that this phrase comes to rest on, or coincides with, a hole.[3] Here, the "hole" Wittgenstein speaks of is presumably a gap in the, or our, colour system. And this may be another way of saying that no space is provided for in our system to accommodate luish yellow. But in saying this we keep moving within the confines of a cluster of very closely related metaphors. These metaphors may support and clarify each other as soon as one has grasped the point. But in order to grasp the point, we may need a few reminders that will assist us in trying to find our way among Wittgenstein's images.

Before getting down to mentioning a few characteristic features that we may helpfully invoke in our attempts to get a clearer grasp of the relevant notion of a system, I just want to allude to a couple of lines which can be found on the same page of Wittgenstein's notebook from which my title quotation is taken. Here, he asks himself whether there is something arbitrary about the system of colours. And he replies by saying "Yes and no. It is akin both to what is arbitrary and to what is non-arbitrary".[4] And this, I suppose, we may take as a first hint indicating that in Wittgenstein's eyes the relevant notion of a system is not a particularly clear-cut affair. That is, we will be well-advised to expect no more than a clarification which, while it may help us to see things more distinctly, does not provide us with neatly drawn outlines and streamlined models.

In attempting to sketch a few characteristic features of a system of the relevant kind I shall basically rely on passages from manuscripts written at roughly the same time as my title quotation. This material comes from several of Wittgenstein's large-sized notebooks written between 1946 and spring 1949. Later, I shall also draw on passages drafted during the last two years of his life. So, this latter material is taken from the same notebooks that were used by the editors of the remarks published as *On Cer-*

---

3 "'Bläulichgelb' fällt auf ein Loch" (137:6a).
4 RPP II, 427.

tainty, *Remarks on Colour* and the second volume of *Last Writings on the Philosophy of Psychology* as well as the last pages of *Culture and Value*.

2 Principles of collection

The first and quite obvious characteristic that I want to mention is a system's function as providing a principle, or principles, of ordering or organizing or arranging things. Here, there are a great number of phenomena that can be seen to shade into each other but which might all the same be distinguished. Thus there are principles of organization, as they are called in the terminology of Gestalt psychology, which could be grouped under our present heading. But owing to *their* connections with other kinds of features of systems, these principles of Gestalt organization will be treated as falling under a separate category. On the other hand, I shall for simplicity's sake count what I want to call "principles of collection" as elements of the first class. By this I mean various kinds of criteria that help us to group things in a perspicuous manner. Here, Wittgenstein likes to speak of ways of putting things together. "*So* will ich die Dinge zusammennehmen", he exclaims at one point:[5] "*This* is the way in which I want to put them together." And here saying that I *want* to put them together in this particular way serves to highlight the fact alluded to in the passage following my title quotation: that *our* nature, our nature as human beings, should be taken into account when it comes to considering our systems. On the other hand (and this may explain some of Wittgenstein's hesitation – his question "How should one put it?"[6]), finding the right, and hence a non-misleading way of bringing in human inclinations, interests and needs is a delicate matter. Later, I shall briefly return to this question.

The following is a characteristic passage explaining Wittgenstein's idea of creating order by means of a system:

> Imagine the world as a beautifully ordered one. There is a drawer for each and every thing. All of it is neat and clean. There is only one thing that does not fit into any drawer. Now the one emotion we feel is *this:* "If only this thing didn't exist! It spoils the marvellous order of things." One's attitude towards this thing is purely antagonistic. One doesn't say: "This object too will have its place in our world." Rather, one will say: "It is filth, vermin, a weed."
> Once we have our nice and neat filing-cabinet, and there is only one thing that doesn't fit, it would be our dearest wish simply to get rid of it. But if someone supplies us with a different system of drawers, and the thing which used to be homeless finds its place, our attitude towards this thing changes completely.

---

5 137:7a. More precisely, these words are quoted: "So will ich die Dinge zusammennehmen!' könnte man sagen."
6 RPP II, 426.

> A system of drawers means an *habit*. An habitual gesture. As soon as we have a new system, we acquire a new habit, a new technique. ("Man is a creature of habit.")[7]

Another example Wittgenstein mentions is that of a man who attempts to give a description of a place but finds himself unable to do so. All of a sudden he does not find his way about: *Er kennt sich nicht mehr aus*. What the man lacks is a proper system. All the systems that come to his mind are inadequate. He thought he was in a well-laidout garden, but to his surprise he finds himself in a complete wilderness. Wittgenstein writes: "Even though there are rules that occur to him, reality presents him with nothing but exceptions."[8] In other words, the man without a useful system does not find his way about; he does not have a means of getting his bearings.

3 Open vs. closed systems

The second feature of systems that I want to mention is the fact that there is a certain way of dividing them into closed and open systems. Evidently, there is a connection between the distinction itself and certain questions that Wittgenstein found especially exciting and important. Examples are the problem of explaining the difference between the "'and so on' of laziness" and the "'and so on' of infinity"[9] or the insight that giving reasons makes sense only if, in a specifiable context, there is a point in speaking of a last, or ultimate, reason – a reason which gives us peace or at least renders further questions pointless. There is more than one way in which the relevant distinction is connected with our systems of number and colour. Wittgenstein explains it by means of the following story:

> Counting, calculating etc., in a closed system in the way a tune is a closed system. The people count with the aid of the notes of a special tune; at the end of the tune the series of numbers comes to an end. – Am I to say: Of course there are further numbers as well, only these people don't know them? Or am I to say: There is also another way of counting – namely what *we* do – and this these people do not know (do not do).[10]

One way in which this story is connected with the question about systems of colour and number is this: that if we have a system of, for example, four primary colours plus black and white, we in this sense operate with a closed system of colours. On

---

[7] 132:57-9.
[8] 132:201-2.
[9] Think, for example, of the difference between a notational abbreviation signifying a musical repeat, on the one hand, and π as signifying its expansion (3,14159 ...), on the other.
[10] RPP I, 647. (Volume I of RPP: ed. by G. E. M. Anscombe and G. H. von Wright, trans. by G. E. M. Anscombe [Oxford: Blackwell, 1980].)

the other hand, we may cherish the idea that there are indefinitely many shades of colour between certain fixed points called "primary" colours. And to the extent the number of these shades really is understood as unbounded, we seem to be using an open system. As regards number, Wittgenstein has just described a closed system. The often-mentioned primitive system of counting up to five and no further would be another system of this kind. Our everyday system of counting, however, and most other mathematical systems in our repertoire are open ones. So, in the light of these considerations our original question can be extended if we get curious and wish to find out whether openness and closure are in any way connected with, or dependent on, our nature as operators of these systems or otherwise involved in the nature of things.

But as I have indicated, closure is not necessarily dependent on the finitude of a series or sequence. There may be no way of getting outside the system for the simple reason that the system in question is such that no attempt at stepping outside would make it easier for us to grasp what needs to be done inside the system. Wittgenstein explains this by referring us to a situation of the classical rule-following type:

> He must go on like this *without a reason*. Not, however, because he cannot yet grasp the reason but because – in *this* system – there is no reason. ("The chain of reasons comes to an end.") And the *like this* (in "go on like this") is signified by a number, a value. For at *this* level the expression of the rule is explained by the value, not the value by the rule.
> For just where one says "But don't you *see* ...!" the rule is no use, it is what is explained, not what does the explaining.[11]

4 Problems

At first glance, at any rate, the third point I want to mention may seem a bit far-fetched and difficult to find a place for. I hope to be able to show that it is of a certain importance, in particular in regard to various remarks Wittgenstein makes on questions about colour. One difficulty about this point is linguistic; for there is no direct and unproblematic way of translating the relevant words into English. The central word occurs in the following passage, which may also serve to bring out my difficulty. The following is Miss Anscombe's rendering of this passage:

> Translating from one language into another is a mathematical task and the translation of a lyrical poem (for example) into a foreign language is quite analogous to a mathematical *problem*. For it is certainly possible to formulate the problem "How is this joke (e.g.) to be translated by a joke in the other language?" – i.e. how is it to be replaced; and the problem may also be solved; but there wasn't [a technique,] a method, a system, belonging to the solution of it.[12]

---

11 RPP II, 404-5.
12 RPP I, 778. The variant formulation "a technique" is not included by the editors.

In a way, this translation is fine but probably it would have been more idiomatic to translate Wittgenstein's word "Aufgabe", which is here rendered as "task", by "problem". What happens in the original is that Wittgenstein then goes on to use the German word "Problem" (as he often does) in the sense of the English word "problem". In this sense, "2 + 2 = ?" is a mathematical *problem*, but in German it would be a *mathematische Aufgabe*. So what Wittgenstein does in the quoted passage is that he uses an Anglicism in emphasizing, first, the *mathematical* character of the problem of translating poetry or jokes and, second, the possibility of solving such a problem if it is given within a system which can provide us with a technique for finding a solution. The chief point of the remark is that it draws our attention to similarities between certain problems of translation and mathematical problems. The matter is not obvious – and our attention needs to be drawn to it – because the system in terms of which a technique for solving the problem can be used is not self-evident. I take it that the system needs to be added or constructed. So what Wittgenstein may be saying would amount to something like this: "Look at your problem of translation as if it were a mathematical one; then you will supply your own system in terms of which the problem will be easier to define and hence easier to address and to solve, for the same terms may yield a technique whose application will allow you to answer your original problem which, through the whole process, has turned into a different one, viz. a kind of mathematical problem."[13]

No doubt these considerations will prove helpful in looking at certain passages from Wittgenstein's *Remarks on Colour*. The sort of passages I have in mind is neatly exemplified by the following remark, which I shall quote in the translation by Linda McAlister and Margarete Schättle:

> 110. If you are not clear about the role of logic in colour concepts, begin with the simple case of, e.g. a yellowish red. This exists, no one doubts that. How do I learn the use of the word "yellowish"? Through language-games in which, for example, things are put in a certain order.
> Thus I can learn, in agreement with other people, to recognize yellowish and still more yellowish red, green, brown and white.
> In the course of this I learn to proceed independently just as I do in arithmetic. One person may react to the order to find a yellowish blue by producing a blue-green, another may not understand the order. What does this depend upon?

The translators have tried to circumvent the difficulty of finding a proper rendering of Wittgenstein's word "Aufgabe" by removing the analogy with the mathematical case: instead of trying to find the solution to a problem, the person in question is said to

---

13 Cf. passages like the following ones: "Du stellst Dir Aufgaben und löst sie dann; wie ein Mathematiker. / Die Aufgabe: non & ne" (137:112a). ("You set yourself problems and then proceed to solve them, just like a mathematician. The problem 'non and ne' [i.e. the problem of distinguishing two different types of negation].") "Viele Knoten entwirren, das ist die Aufgabe des Philosophen" (138:7b). ("To disentangle many kinds of knot – that's the philosopher's problem / task.")

react to a given order. In this particular case, I think this is an unfortunate solution to our problem of translation precisely because it removes or ignores the analogy with mathematics. In this way it also removes the necessity of seeing the problem as articulated within a certain system, which in *its* turn may provide a technique helpful in solving the problem. Another feature which tends to get lost in this translation is a particular gloss one may wish to put on the language-games invoked in the first paragraph of our quotation. These language-games, I gather, should also be seen as ways of setting and solving problems comparable to arithmetical ones. That is, they should be seen as problems within a specific system. This is not intended as an improvement on our actual means of teaching and learning expressions like "yellowish"; it is meant to bring out their unobvious systematic character in order to make us grasp – by way of analogy – what it is that needs to be learned. The system as well as the rules that might be quoted to spell out what we grasp are in the nature of what Wittgenstein calls a *Hilfskonstruktion*[14] – an auxiliary construction line, a line which is drawn for the purpose of constructing a complicated geometric figure and which may be erased or ignored once the figure has successfully been constructed.

Similar considerations would apply to other passages in the published *Remarks on Colour* where Wittgenstein uses the word "Aufgabe". Here I am thinking of passages like the following:

> 47. What does it mean to say, "Brown is akin to yellow?"
> 48. Does it mean that the task of choosing a somewhat brownish yellow would be readily understood? […][15]

Here, the "task of choosing" is, I think, the quasi-mathematical problem of picking out a certain shade of colour. It is a quasi-mathematical one because the order in which these shades are given depends on our colour system and its attendant techniques of organizing these shades. – Another relevant passage is the following:

> 111. *I* say blue-green contains *no* yellow: if someone else claims that it certainly does contain yellow, who's right? How can we check? Is there only a verbal difference between us? – Won't the one recognize a pure green that tends neither toward blue nor toward yellow? And of what use is this? In what language-games can it be used? – He will at least be able to respond to the command to pick out the green things that contain *no* yellow, and those that contain *no* blue. And this constitutes the demarcation point 'green', which the other does not know.

---

14 See, for example, 137:69b: "Ich mache immer Hilfskonstruktionen, die am Ende aus der Betrachtung herausfallen sollen" ("I keep producing auxiliary constructions that are ultimately meant to be left out of consideration"); 138:5a and 22a; 144:95 (= PPF 318); Zettel, 528. (PPF = Philosophy of Psychology – a Fragment = Part II of the first three editions of Wittgenstein's Philosophical Investigations; for the fourth edition, see below, note 20. Zettel was edited by G. E. M. Anscombe and G. H. von Wright, trans. by G. E. M. Anscombe [Oxford: Blackwell, 1967].)

15 These and the following quotations are taken from part III of Wittgenstein, Remarks on Colour, ed. by G. E. M. Anscombe, trans. by Linda L. McAlister and Margarete Schättle, Oxford: Blackwell, 1977.

The command to pick out green things of a particular kind is, again, an *Aufgabe*. And I'm sure our understanding of this passage would be improved if we saw that Wittgenstein is talking about a quasi-mathematical problem, whose very existence and solution depend on the employment of one system rather than another one. The "demarcation point" Wittgenstein speaks of can fulfil the role attributed to it by one player of certain language-games only if the relevant system is the one in whose terms these language-games are understood and discussed.

To mention yet another example of Wittgenstein's use of "Aufgabe". In the context of a longish remark he says that "There is a more or less bluish (or yellowish) green and someone may be told to mix a green less yellow (or blue) than a given [...] one". Wittgenstein, however, does not speak of someone's being told to do this; he says that there is the *Aufgabe*, the problem, of mixing such colours. Again, it seems quite important to see that the order to mix something of such and such a kind is not to be seen as a purely practical task which may be fulfilled satisfactorily by trying things out in random fashion. No, we are dealing with a problem which needs to be seen in systematic terms; and we are talking about quantities which can, at least theoretically, be measured against a continuous line constituted by what Wittgenstein calls the "coloured path from blue to yellow" on which green figures as "one special waystation".[16] In other words, the existence as well as the solution of the intended kind of problem of mixing colours involves the acceptance of a particular colour system as opposed to another one. Part of the meaning of this insight might be captured by saying that the kind of mixing of colours intended in speaking of such problems is not a purely practical, but an highly theoretical or quasi-mathematical kind of activity.

5 Aspects

There is another use of the word "system" which should be pointed out. On the one hand, it is connected with things that have already been said. On the other, it relates to a topic which as yet has not been touched on. The topic I mean is that of aspect seeing, of seeing something as something or of seeing something as something else. Wittgenstein's discussions of figures of the duck-rabbit kind are a well-known instance of this.

The notion of a system enters these discussions at a point where Wittgenstein expresses his dissatisfaction with the Gestalt psychologists' tendency to explain phenomena of the relevant kind in basically physiological terms. Wittgenstein approaches

---

16 40. Perhaps 'colour path'(rather than 'coloured path') would be a better translation. Cf. 41: 'direkter Farbenweg' – in the published English version, translation of Wittgenstein's peculiar phrase has been avoided.

these phenomena by asking what "system of modifications" would be sufficient to accommodate the generating of certain aspects. Would a "system of shapes and colours" do? I take it that according to Wittgenstein a psychologist like Köhler (who is quoted in this context) would try to show that such a system can be enough. This would mean that physiology plus colour and shape terms suffice to describe and and explain what we see when we see a certain figure now as a cross in a circle and now as a circle with a cross-shaped gap. The colours, for example, would be ordered in accordance with one of the usual models and thus form an organized system that would function as a kind of yardstick against which the phenomena to be described can be laid and measured.

My impression is that Wittgenstein is sceptical about the Gestalt psychologist's claim. He refers to a detail of the duck-rabbit figure and asks: "When I see the dot as an eye which is looking in *this* direction – what system of modification does that fit into?"[17] Evidently, the suggestion is that a system of the kind proposed by Köhler and other psychologists of his school would not be enough. It would not even be enough to *describe* the phenomenon intended because capturing the idea of looking in a certain direction requires more than the austere system offered by the psychologist. Not only answering possible questions about these phenomena but even articulating such questions requires more context and a richer conceptual apparatus than is on offer.

As Wittgenstein says, coming to see something that appears to have no structure as a pattern of a certain kind is like coming to see a system where there used to be only irregularity.[18] And what helps me to come to see this system is another system – a "system of modifications" (as he calls it) in terms of which I may be able to find my way about in the conceptual space spread out by this second system.[19] Perhaps the most striking formulation of this notion is Wittgenstein's repeated observation that only in the context of a face can eyes or lips smile.[20] Analogously, the dot in the duck-rabbit figure can be said to look in a certain direction only if ducks or rabbits – and hence much more than mere systems of colour and shape – enter into the discussion.

---

17 RPP I, 1116.

18 135:70v-71r: 'Ich schaue eine Tapete an. Ihr Muster ist zuerst für mich ein regelloses Gewirr von Flecken; nach einer kurzen Prüfung kenne ich mich aus; es ist ein System." ('I look at a stretch of wallpaper. At first, for me its pattern is nothing more than an irregular tangle of marks. After a brief examination I have come to know my way around. It forms a system.")

19 RPP I, 1122

20 Last Writings on the Philosophy of Psychology, Volume I (ed. by G. H. von Wright and Heikki Nyman, trans. by C. G. Luckhardt and M. A. E. Aue [Oxford: Blackwell: 1982]), 860; Philosophical Investigations (4th edition by P. M. S. Hacker and Joachim Schulte, trans. by G. E. M. Anscombe, P. M. S. Hacker and Joachim Schulte [Malden, MA, and Oxford: Wiley-Blackwell, 2009], 583.

6 Language-games: our systems, our nature

The remarks Wittgenstein wrote in his last years contain a number of further uses of the term "system" that would be relevant to our considerations. Perhaps the most important one is a use well-known from the published remarks *On Certainty*,[21] where our world-picture (as Wittgenstein calls it) is compared to mythological and religious systems. Some notions which can be seen to be particularly important features of such mythological or religious systems can then be recognized as characteristics of our own world-picture. Examples are special forms of commitment and the fact that a change of world-picture involves a kind of conversion analogous to a change of religious faith. Another crucial feature is the connection between these systems and our traditional, ingrained ways of acting and habits of thought. Of course, this is an aspect of the matter which recalls our original question whether colour and number systems reside in *our* nature or in the nature of things. After all, it is obvious that these ways of acting and habits of thought have something to do with our nature, at the very least in the sense of being dependent on our nature inasmuch as they come naturally to us.

In his late manuscripts Wittgenstein keeps raising questions about the general dependence of our concepts on our way of living, our needs and desires. On the one hand, he tends to emphasize that these concepts are instruments of language; on the other hand, he does not want to claim that it is all a matter of practical versus unpractical. He points out that concepts are principles of organization, or collection, but at the same time he wishes to make it clear that in virtue of their power as tools for arranging things in a surveyable order they are eminently practical devices. He keeps raising questions like whether our ways of counting or of using colour words satisfy our needs; and he keeps replying that this is not a matter of choice: that this is just what we do and that it forms part of the "natural history of mankind" (as he likes to put it).[22] He also keeps repeating that in this context talking about needs, instruments, practice and life is not a matter of causality: these notions are not brought in to help us give causal explanations of why we have these rather than those other concepts.

As a matter of fact, part of the point of Wittgenstein's remarks may be that there is a general difficulty in this way of speaking – I mean in using phrases like "these rather than those other concepts". But it is a difficulty which it is not easy to account for. Let's look at one of Wittgenstein's own examples involving colour and number systems. As he reminds us, our number system may be "connected with the number of our fingers". And if this is so, "why shouldn't our system of colours be connected with the specific ways in which they occur"? Thus it might be the case that "a colour

---

[21] See my paper "Within a System" (Danièle Moyal-Sharrock and William H. Brenner [eds.], Readings of Wittgenstein's On Certainty [Houndmills: Palgrave-Macmillan, 2005], pp. 59-75.
[22] 137:60b ff.

occurs only in gradual transition into another one"; or that "colours always occur in the sequence of colours in the rainbow".[23] And he goes on to say that a world could be such that all animals have six legs and throw six young, and that people only use six-footed metres. If they also used a base-6 numeral system, we would somehow connect all these facts, and it might even be the case that their arithmetic expressed their abhorrence for all exceptions from the rule. None of this would come as a total surprise, because – to quote from the same manuscript – "as regards a different kind of world, the use of different instruments of language would appear natural".[24] That is, it would appear natural to us if we could identify other features corresponding to the kind just exemplified. In other words, even if certain phenomena do not agree with what appears natural to us, they may none the less seem natural if we are able to see the system behind them.

As Wittgenstein says, we are inclined to explain the existence of certain concepts by referring to our need for them. Presumably, the fact that we *are* so inclined depends on our wish to find causal explanations: we want to discover general laws governing the actual development of language and our conceptual system. Wittgenstein's comment is worth quoting. He says that this need, i.e. the need by reference to which we want to explain our possession of these concepts, is "merely the need for [or, perhaps, the hankering after] a certain way of life which comprises the use of this concept". And he continues: "By pointing out the language-game one shows the connection between language and life. That is, one shows, not a causal connection, but the way language is interwoven with other things that happen in life."[25]

Maybe this way of summarizing and rendering Wittgenstein's remarks is a bit of a simplification, perhaps even somewhat partial. But I think it is roughly right in showing that Wittgenstein does not want to give a straightforward answer to his own question: whether our systems, specifically our number and colour systems, reside in *our* nature or in the nature of things. He does say that they do *not* rest on the nature of numbers or colours. But it looks as if he did not want to admit that they clearly reside in our nature or in the nature of things different from numbers and colours themselves. If I am allowed to speculate, I should say that he would not really mind conceding that our nature as well as the nature of things have something to do with the existence of these systems. But he would not want our nature and the nature of things to enter the equation *separately:* we are invited to see that there is a point where several factors meet in a way which deprives them of their mutual independence and makes them

---

23 RPP II, 658 (cf. 137:61a).
24 137:61b.
25 137:61b. ("Wir wollen uns Begriffe aus dem Bedürfnis nach ihnen erklären. Aber eigentlich wäre das nur das Bedürfnis nach einer bestimmten Lebensweise, welche die Verwendung des Begriffs in sich schließt. / Indem man das Sprachspiel aufzeigt, zeigt man die Verbindung der Sprache mit dem Leben. D.h. die Verwebung der Sprache mit den andern Lebensvorgängen, nicht einen kausalen Zusammenhang.")

appear as integral parts of an organic unity. This unity is called the language-game. And the language-game is something of which one can give a demonstration; one can perform it like a play. And by doing so one may succeed in exhibiting the way life and language are interwoven. Of course, this can work only if by "life" one does not mean an hotchpotch of undifferentiated reality but something that has a kind of structure, an organization. In this sense, life as well as language lend themselves to being represented through performances of language-plays. They are highly stylized ingredients in an highly stylized process of acting. Perhaps one may say that to the extent the system behind our nature and the system behind life can be seen to be in harmony, we need not decide whether the systems of colour and number reside in one rather than the other. And if we can manage to accept it that there is not much difference between the nature of things and life itself, our original question answers itself, as it were: we have a system of colour just as we have a system of numbers; and this means that our system, our nature, and the system of life, the nature of things, are in agreement to the extent that they can figure in language-games (or language-plays) employing these systems of colour and number. Or to put it in yet another way: our original question was never meant as a serious question. Reflection on it was meant to show that, if we have functioning systems of colour and number, *they* will be as informative about us as inquiring into *our* nature will be revealing about them.

Richard Heinrich
Green and Orange – Colour and Space in Wittgenstein

1 Introductory remarks

I will refer mainly to remarks by Wittgenstein from around 1930. But in some sense my talk is focused on a few sentences from the beginning paragraphs of Philipp Otto Runge's "Colour Sphere" from 1810,[1] where he describes the first steps in the construction of the sphere. What I find fascinating in these passages – and what I want to give an impression of – is the methodological chaos behind the seemingly unproblematic model. Many of the tensions and inconsistencies to be found there reappear in Wittgenstein's thinking about colour. Some of them are recognized by him as characteristic or even defining problems in the field, in the sense of Jonathan Westphal's "puzzle questions" (Westphal 1987: 3) like, for instance: "Why can't something be of a reddish green?" In the following I will not so much be concerned with particular such problems, but rather with the consequences they may be supposed to have for the very idea of a model for the colour system, in the sense of a "Farbraumdarstellung", a spatial representation or a geometrical representation.

From the period of the TLP onward "colour" is a constant subject in Wittgenstein's philosophy, and given that some of the most intriguing problems he thought about had to do with the structure of colour space, it could be found amazing that he more or less faithfully held on to one model of colour representation throughout, the double pyramid of the type of Alois Hoefler's colour solid (cf. Wilde 2002: 284). This cannot be taken as an indication that Wittgenstein's views about colour representation did not change – but at least one point of lasting significance for Wittgenstein can indeed be brought out by contrasting Hoefler-type models with, say, a colour sphere. What might have changed over time are Wittgenstein's views regarding the general meaning of spatiality in this kind of representations, below the level of adherence to any particular model, and in particular the role to be accorded in this context to "geometry". Let me say a few words as to what I mean by this.

Different models for colour space serve different purposes and, as Westphal put it: "[…] they record different type of data […and] are also governed by different types of concepts, some physical, some psychological." (Westphal 1987: 95f). Sometimes these differences are immediately reflected in the design of the model, but that is not neces-

---
[1] Quotations from Runge's "Colour Sphere" are taken from the translation by Rolf G. Kuehni.

sarily so and when it is the case, it does not necessarily catch the eye. When Wittgenstein says something like the following: "The colour octahedron is used in psychology to represent the scheme of colours. But it is really a part of grammar, not of psychology" (LWL, p. 8), he does not ask himself how a model (or representation) should look like which really served psychological purposes. And when he reflects upon the differences between geometrical designs, like in remarks 221ff. of *Philosophical Remarks* (where he considers the substitution of the double pyramid by a double cone) his aim is not to show that the one is suited for a practical purpose the other cannot fulfil; he rather uses this reflection to articulate a theoretical issue within the field of grammatical analysis, namely the specification of pure colours. (On the cone there is "only a between and red appears on it between blue-red and orange in the same sense as that in which blue-red lies between blue and red".) In fact Wittgenstein says in this passage that short of topological differences all the various models could somehow be made equivalent.

When thinking about models for colour space Wittgenstein was also systematically engaged in questions which concern his use of the concept of space as such and in contexts which at first sight might seem quite remote, like the structure of logical space or the issue of metaphorical space. Here Andrew Lugg's assumption becomes relevant that before the time of RC Wittgenstein "accorded colour and colour language no special treatment and mentioned them only in the course of discussing more general philosophical matters"[2]. I find it worth asking whether there is more to these "mentions" than just an illustrating function in the service of this or that issue coming up on a higher level of abstraction. The development of a certain branch of Wittgenstein's terminology, from the conception of logical space in the TLP via the idea of the "determination of a coordinate" around 1930 to the notion of "perspicuous representation" seems to me to justify interest in such a massive body of thought as the reflections on models for colour space make up in their totality from the TLP onward. (This does not mean that some of the matters discussed could not also have been raised with respect to sounds, numbers or visual space. But I suspect that these topics stand in relations of mutual supplementation rather than possible substitution of one by the other.)

---

[2] The quote is taken from the English original of an article titled "Wittgenstein on Colour. 1949/1950." published in 2009 in the collection *Sense and Nonsense. Wittgenstein and the critique of language.* which was edited by Dans Carlos Moya and Luigi Perissinotto.

2 Wittgenstein around 1930

When Wittgenstein distanced himself from his early atomism he observed (cf. PR, 83) that already in the TLP the remark that a coloured body is in a colour-space (cf. TLP, 2.0131) could have prompted the idea that between elementary propositions there exist logical dependencies – what is now captured by the concept of a coordinate. The reason he had not seen this was his failure to realize the tensions between his "logical space" on the one hand, and spaces corresponding to special forms of representation (like colour) on the other hand. A remark like TLP, 2.182: "Every picture is also a logical picture. (On the other hand, for example, not every picture is spatial.)", is misleading insofar as it subliminally suggests a seamless integration of two kinds of relationship which are in fact incompatible: In the logical space of the TLP every elementary proposition is as it were a separate dimension, whereas in colour-space every proposition marks a position in one and the same dimension as infinitely many others. It is instructive to read one passage from a conversation with Schlick and Waismann from 2nd January 1930, where Wittgenstein reflects more closely on his TLP-conception of elementary propositions (cf. WVC, p. 73f). He cites two ideas as characteristic: Mutual independence, which he now recognizes as a "complete error"; and the idea that elementary propositions are immediate connections of objects. To this second view he sees himself still committed, and he accentuates it in way which links it to "the very first philosophical remark of Wittgenstein's that has been preserved." (McGuinness 2002: 104):[3] "Immediate connection" here means above all that the elementary propositions do not contain logical constants like "not", "and", "or", "if – then".

But there is a difference: For in the conversations with Schlick and Waismann Wittgenstein's main topic is a kind of proposition "von der ich früher keine Ahnung hatte" – "a kind of proposition of which I used to have no idea and which corresponds roughly to what I want to call an incomplete picture" (WVC, p. 39): Elementary propositions in which some determinations are "left out" and which can contain variables. He cannot claim of this kind of propositions that they consist entirely of simple signs, but he can still maintain that they do not contain logical constants of any sort. They are internally structured in a way which makes them dependent on one another, but not truth-functionally. There is one quotation from another conversation at about the same time where he puts the matter in almost shockingly direct fashion: "The logical structure of elementary propositions need not have the slightest similarity with the logical structure of the propositions." (WVC, p. 42)

This is the point of origin of the word "grammar" in its systematic meaning for Wittgenstein, and it is of utmost importance to see that this concept does not simply

---

3 The reference is to Wittgenstein's letter to Bertrand Russell from 22nd June 1912, quoted in McGuinness 2012: 30.

explain itself, or is simply explained by its opposition to "logic"; one has to look attentively at the field of problems and alternatives where it is initially positioned. Of course, the first thing which meets the eye (and could already have met the eye of the author of the *Tractatus logico-philosophicus*) is the existence of an inferential rapport between elementary propositions by virtue of their special form; but that only poses the further question of the relationship between this kind of dependency on the one hand, and what has hitherto been called a logical relationship. And here we see that Wittgenstein could bring out the essential point as well by using the word "syntax" as by using the word "grammar". In conversation with Schlick and Waismann on 2nd January 1930 he says: "What was wrong about my conception was that I believed that the syntax of logical constants could be laid down without paying attention to the inner connection of propositions [...] Rather, the rules for the logical constants form only a part of a more comprehensive syntax about which I did not yet know anything at that time" (WVC, p. 74). In remark 83 of *Philosophical Remarks* he says practically the same using the word "grammar": "The rules for 'and', 'or', 'not' etc., that I represented in my binary T-F-notation, are just a part of the grammar of these words, but not the whole" (PR, 83). I could quote numerous passages from the conversations where Wittgenstein expresses characteristic traits of his notion of grammar using the word "syntax", e.g.: "Language is already perfectly ordered. The difficulty is just to make the syntax simple and perspicuous" (WVC, p. 46). The point is always that inferential relationships of the truth-functional type (including probabilities) do not make up a system more basic than relationships of the "spatial" or "coordinative" type, but are part of that "enlarged syntax" – they are a sort of extraction from the more comprehensive grammar, a projection of certain features of it onto a model which is useful for certain well defined purposes (logicians purposes, as he says once). Already in the TLP Wittgenstein held "that one cannot foresee the form of elementary propositions" (WVC, p. 42; cf. TLP, 5.55, 5.556, 5.5571), but now this is to be understood in a much more radical sense: Think of the possibility that the form of certain elementary propositions might be determined by real numbers – impossible to foresee what may confront us here.

Now if we take the remark from TLP, 2.0131: "The speck in the visual field need not be red, but it must have a colour" as implying that also something like "This spot on my forearm is red" is an ordinary proposition, the reflections sketched in the foregoing make it clear that this cannot without further qualifications be called an elementary proposition. We cannot right away produce the equations which have to obtain between certain particles so that the proposition comes out true. We do not know which is the ultimate level of analysis here to target at. If we take, on the other hand, a proposition like: "If the spot is not red, then it must certainly be blue or yellow or green or some other colour" – what role does the word "colour" play here? (What do we understand when we understand the words "some other colour"?) We cannot be certain that seen from the first perspective (possible reduction to physical equations) there will be anything at all corresponding to it. (And there is some evidence that there really won't be.) But nonetheless here it seems to stand for or even guarantee the kind of internal relationship or grammatical structure which Wittgenstein had in mind with his "enlarged syntax". If we grant ourselves a preliminary use of the word "colour", and set out on this level of granularity to describe the relationships it covers – especially what is meant by the phrase "some other colour" –, then we will find ourselves on an intermediate level between "real" elementary propositions (where "logical multiplicity is not depicted by subject and predicate or by relations, but e.g. by physical equations" (WVC, p. 43)) on the one hand, and logical relationships in the sense of truth-functions on the other hand.

And to spell out this kind of relationships we use a model – in the case of colours a geometrical representation like the octahedron: "The colour octahedron [...] is really a part of grammar [...] It tells us what we can do: we can speak of a greenish blue but not of a greenish red etc. [...] grammar is not entirely a matter of arbitrary choice. It must enable us to express the multiplicity of facts." (LWL, p. 8).

When we say of such a model that it gives us a representation of the colour space, the profile of the word "space" has been sharpened in contrast to the colour space (Farbraum) of remark 2.0131 in the *Tractatus logico-philosophicus*. There it was some sort of re-import of the highly abstract notion of a "space of possible states of affairs (or atomic facts)" and as such did not imply any original meaning of spatiality with regard to colours. But when Wittgenstein says of the colour octahedron that it "tells us what we can do" he means that it lets us see something by virtue of being a spatial arrangement. This connection between seeing and space is weak, anyhow, and there is a wide range of meanings for the "spatial arrangement", from a child's arranging of colour pencils to more ambitious models. Philipp Otto Runge set out to construct his colour sphere as a geometrical model in a rather strong sense. His paradigm are the scientific foundations of drawing, as given by the geometrical laws of form, proportion, and perspective. When we say of such a model that it "tells us something", we somehow appeal to a tacit understanding of a language the model itself speaks, the language of geometry. And the connections between this language, the visible arrangement and what it is a model for are not always straightforward.

3 Runge

On a programmatic level Runge definitely sets out to construct a geometrical model:

> "There have been frequent efforts, even though only at the level of attempts, to represent the relationships between all mixtures in tabular form. The geometric figure that is to represent the complete connection of all relationships cannot be arbitrary. Rather, it must represent the relationship itself and must necessarily arise out of the natural inclination or disinclination between the colors, as expressed by the elements." (Runge 1810: 9)

The method he proposes is: Certain basic elements are given a geometrical interpretation and then conclusions are drawn from the ensuing relationships. The most abstract concepts he works with are "pure colour" and "mixture", and we could as well take a quotation from Wittgenstein's conversations with Schlick and Waismann as a statement of his program: "Whatever colour I see, I can represent each of them by mentioning the four elementary colours red, yellow, blue, green, and adding how this particular colour is to be generated from the elementary colours" (WVC, p. 42). The only additional clause being the restriction of "production" to "mixture" (which may also be responsible for the difference in the number of prime colours). Accordingly, Runge takes red, blue and yellow as pure colours, completely idealized, and without any admixture of the respective others. They are conceived in analogy with the geometrical point:

> "[S]uch posited colors, ideal and free of any admixture, can be represented by dimensionless, geometric points. And, because the quality of each of the three colors is of a completely individual nature, different from all qualitative aspects of the other two, I equate the distance between them. As a result, the three points representing blue, yellow, and red, their distance expressed by lines of equal length, form an equilateral triangle." (Runge 1810: 20)

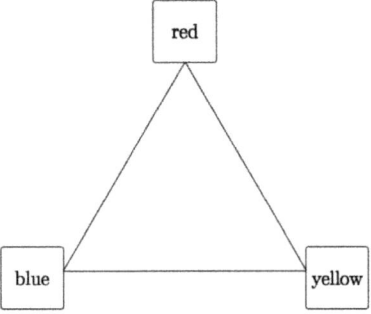

So in fact the three pure colours are geometrically indistinguishable among themselves, as well as the three lines joining them. The only geometrical difference is between points and lines, which are objects of different kinds. And the relationship between, on the one hand, red, yellow, and blue (Wittgenstein added green), and "any other colour" on the other hand should now be explained via the relationship between the points and the lines respectively in the triangle. The vital point here is that the spelling out of that relationship alone is what can give content to the generic concept "colour" in a phrase like "if it is not red, than it must be either yellow or blue or green or some other colour". Now, to treat the lines themselves as colours is not possible right away; the word colour has to be re-interpreted – a possible general use of the word "colour" has still to be constructed, to be established from the given primitives "pure colour" and "mixture"; Runge's first idea here is to relate the lines to the points via the concept of "inclination" or, more general, "movement". The lines are inclined towards the points. Already here one can sense a problem: Inclination and movement are not immediately recognizable as geometrical concepts; and even if an appropriate interpretation be granted in principle, the concept of inclination in itself does not yield a sufficiently precise picture. It requires, in addition, some sort of law for the specific way the differences between parts of the lines are ordered. I think that Runge himself felt a problem here and that he reacted with two different ideas. One consists in adding still another interpretation of the line, on top of its geometrical identity and the kinetic interpretation as movement already given, a dynamic interpretation: The points act on the lines with certain forces. (This definitely shows that he has lost control over his geometrical model and has to fall back on sub-models to lend it expressiveness). Independently from that he seems to have hit on an idea which gave him an opportunity to combine two problems into one solution. As no feature or element has been introduced so far which would geometrically distinguish any directions on the triangle, it can seem natural to assume that each line is inclined towards the two points it joins in the same degree; and this goes very well with the pre-theoretical knowledge that phenomenologically three additional colours – green, orange and purple – make their appearance on those lines. The phenomenological individuality of these colours (or: so-called colours) in a natural way identifies itself as the center points of the lines.

> "[G]reen will appear as a singular color at the center point of the line BY, equally inclining toward blue and yellow and having the same distance from the two." (Runge 1810: 22)

Runge recognizes the danger lying in the expression "singular colour": It tends to obliterate the difference in kind between pure colours and mixtures:

> "Contrary to the unitary nature of each of the three points B, G, and R, each of the three kinds of mixtures, green, orange, and violet, are multiples, existing in countless grades between the two colors." (Runge 1810: 22).

He is aware that, basically, orange, green and purple do not exist as points but in "countless grades". A proper solution should preserve this difference, and keep it visible in the model. But Runge is content with acknowledging it discursively, in the text, whereas in his diagrams he represents green, purple and orange by points. And once these points exist, a further geometrical entity comes into existence: a second equilateral triangle. First it appears smaller and upside down within the first triangle:

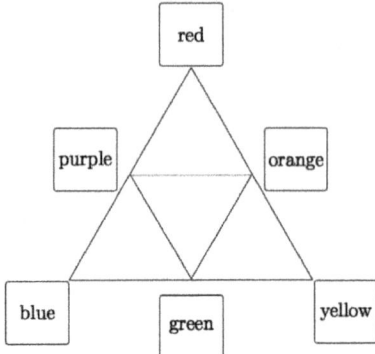

But geometrical considerations, if they are shrewd enough, can act like a fertilizer: If we also take into consideration the two pure colours white and black, and posit them in the same geometrically indistinguishable way as red, blue and yellow, then there seems to be no reason why not also the "singular points" should be in the same distance to black and white – which lets the triangle grow outwards until it has, still upside down, the same size like the first and we have constructed a regular hexagon:

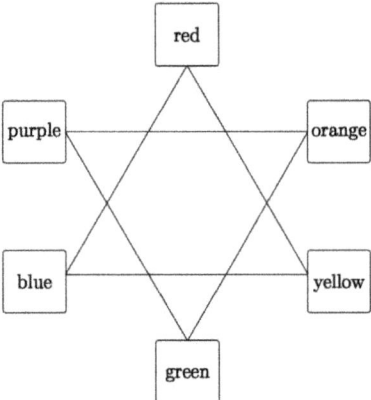

Almost needless to add that now he hastens to acknowledge the existence of the complete colour-circle through these six points – whereby the distinction between pure colour and mixture definitely vanishes from the model. Every point on the circle is a colour.

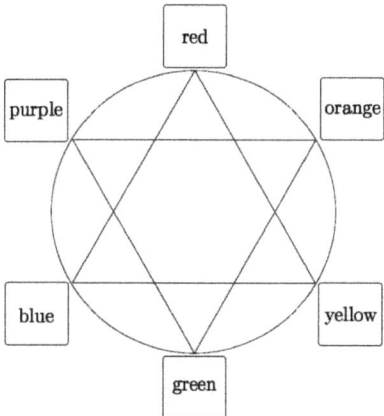

4 Wittgenstein: Conclusion

There are several criticisms possible here from a Wittgensteinian perspective. To take one of the most familiar points (and which gave me the title for my talk) first: Structural equality between the three mixed colours (or "singular points") is an illusion; of course, there is one sense in which mixture on each of these lines alike produces "something new"; but we will have to admit that green lies between blue and yellow in a fundamentally different sense as orange between red and yellow. We can, in the second case, add yellow to red and the red will grow more and more yellow and at some point we will call that orange; but we could as well see it as a very yellow (or yellowish) red, and if we continue on the other side of the center point we will find a reddish (less and less reddish) yellow. Not so with green: There is no blueish yellow (cf. RC I, 9, 14, etc.). Wittgenstein treats the case of yellowish blue as on a par with reddish green. That is a main reason for his preference for the octahedron. But there is a systematically deeper point to that: To say that geometry gives us the model, to say that we just define one or two basic elements, and then geometry does the rest for us, is an illusion. There is a specificity to colour which cannot be traced back solely to the nature of basic entities for which corresponding entities are provided in the model. There is always a tension between – if I may put it that way – colour space and the geometry of colour.

A second point concerns the transition from the triangle via the hexagon to the colour circle: The dissolution of the difference between pure (or primary) colour and mixture. Remark 221 of *Remarks on Colour* once again:

> "[If one replaces the octahedron with a double cone there is] only a *between* on the colour circle, and red appears on it between blue-red and orange in the same sense as that in which blue-red lies between blue and red'." (PR, 221)

I think that Wittgenstein's reasons for preferring the octahedron to a model with three basic colours, a double tetrahedron, are very important; but a far more fundamental question is concerned with his opposition to the double cone. I think that his opinion was that if we let us guide by such a picture, that we then can no longer speak of a model for the grammar of colour, and that in this case the concept of colour would lose its sense.

Wittgenstein's position was: That when we want to treat "red" as a colour (colour as a concept under which "red" falls), and when we are secondly convinced that saying of something that it is red commits us to the existence of a certain form of internal relationship this something entertains with other possible items and colours – then we are also committed to the difference between primary colours and mixtures. Grammatical analysis, what he also called the "logical analysis of the phenomena", has concepts as its subject. If I say "lemonchiffon" is a colour, it can be asked: Which? And I should be able to answer that, if "colour" is a general concept, by giving its relationships to other colours. Two or three more points in conclusion.

If in this sense the grammar of colour has essentially to do with the concept of colour, it does not, on the other hand, cover our whole experience with (and of) colour. Not every use of a colour-word is systematic, not every report of a certain experience of colour is of the kind that it either presupposes or substantiates a stable meaning for a colour-word.

One way to see this is by reference to the materiality of the models for colour-space. The use we make of the colour octahedron is in fact embedded in a very complex scenario. On the one hand the model stands for something else, say, the structure of our colour-language; on the other hand it is something we actually see. To be able "to be told something" by the octahedron one has not only to be engaged in both aspects, but moreover to be committed to a special compromise, a binding together of those two aspects in what one could call a discipline of seeing. Such a discipline always excludes some elements of our actual competence of seeing colours. It is due to that selective function of the model that one can be surprised by something like the fact that transparency acts like an additional dimension in colour space. We then realize that seeing colours cannot be mapped on a program for reading off relationships from a geometrical form, but takes place as a process in space, characterized by relations of "behind" and "between". In this way, for Wittgenstein, the transparency-problem (see the remark on Runge in RC I, 17) showed the interaction between the actual, infinitely complex circumstances of our use of language, and conceptual analysis proper: The challenge space poses for model-space or, put otherwise, the difference between grammatical analysis and "Naturgeschichte" of colour (RC III, 135).

References

Kuehni, R. G.: *Philipp Otto Runge's Color Sphere. A Translation, with Related Materials and an Essay*, Kapitel Available online: Inter-Society Color Council. http://www.iscc.org/pdf/RungeFarben-Kugel.pdf, 2008.

Lugg, A.: El problema del color en Wittgenstein. 1949-1950. In: Moya, C. J. (Ed.), *Sentido y sinsentido. Wittgenstein y la crítica del lenguaje*, p. 117–136, Pre-textos, 2008.

McGuinness, B.: The Grundgedanke of the Tractatus. In: McGuinness, B. (Ed.), *Approaches to Wittgenstein*, p. 103–115, Routledge, 2002.

McGuinness, B. (Ed.): *Wittgenstein in Cambridge. Letters and Documents 1911-1951*. Blackwell, 2012.

Mulligan, K.: Colours, Corners and Complexity: Meinong and Wittgenstein on some Internal Relations. In: Fraassen, B. C. v. & Skyrms, B. & Spohn, W. (Ed.), *Existence and Explanation: Essays in Honor of Karel Lambert*, p. 77–101, Kluwer, 1991.

Runge, P. O. (Ed.): *Farben-Kugel*. Friedrich Perthes, 1810.

Westphal, J.: *Colour: Some Philosophical Problems from Wittgenstein*. Blackwell, 1987.

Wilde, T.: The 4th Dimension – Wittgenstein on Colour and Imagination. In: Kanzian, C./ Quitterer, J./ Runggaldier, E. (Eds.), *Persons: An Interdisciplinary Approach. Papers of the 25th International Wittgenstein Symposium.*, p. 284–286, Austrian Ludwig Wittgenstein Society, 2002.

Gabriele M. Mras
'Propositions About Blue' – Wittgenstein on the Concept of Colour

> Can't we imagine certain people having
> a different geometry of colour than we do? (RC I, 66)

In the *Remarks on Colour* Wittgenstein considers the possibility of other people's "having a different geometry of colour than we do" (RC I, 66). I want to take up the question of *what* thereby is considered to be so – the *geometry of colours*.

These so called geometrical relations are expressed in statements about the relations amongst the different colours. It is my idea that thinking of these relations amongst colour in this way is an expression of the very difficulty that the appeal to the geometry of colours is meant to overcome.

Wittgenstein speaks in what appears to be the simplest and most innocuous ways about our concepts of colours. One can find:

"There is yellow", "There is blue", "There are four primary colours" etc.

but the understanding of these sentences as expressing the geometry of colour requires acceptance of an exhaustive division of all the different colours that there are. This in turn involves a commitment to the impossibility of there being any other colours at all. This cannot give us the understanding we seek of the familiar colours we all recognize.

I think that this is ultimately the reason why Gordon Baker was opposed to the idea of a geometry of colour. Peter Hacker questions Baker's claim as an interpretation of Wittgenstein:

> Baker held that the colour-octahedron is meant to be [...] not an *expression* of the rules for the use of colour words at all. Accordingly, the colour octahedron does not itself belong to grammar [...] (Hacker 2007: 119)

> It seems to me that Baker's late conception of Wittgenstein's methods was unduly influenced by the work he did on the Waismann papers. (Hacker 2007: 94)

Yet neither Baker nor Hacker goes to the heart of what I think is really at stake in Wittgenstein's *Remarks on Colour*. What is at stake is the extent to which our concept of colour allows an understanding of colour terms that is not exhausted by predicating colour terms to objects. This is a serious issue because any other usage of colour terms seems to result in sentences that are ultimately meaningless. If not (meaningless), this different use of colour terms must be independently explained and defended. Such an explanation would require that this usage of colour terms must be connected with the predicative application of colour terms onto objects. Here we have to distinguish between the usage of colour concepts in predicating colours to objects from their use in describing their relations between different colours and the different sentences in which such claims are made.

As an interpretation of Wittgenstein one might be inclined to accept Hacker's point against Baker that a geometry of colour provides criteria for the application of colour terms. Wittgenstein asks *"isn't it precisely the geometry of colours that shows us what we're talking about, i.e. that we are talking about colours?"* (RC III, 86). And he looks upon grammar – as putting together the primary colours –

> The colour octahedron [...] is a perspicuous representation of the grammar of colour words precisely because it "wears the rules of grammar on its face". (PR XXI, 221)

The problem with any claim as that a geometry of colour provides criteria for the application of colour terms is – to put it mildly – that it is not clear *how* such a geometry could provide criteria for the use of colour terms.

How does the colour octahedron manage to "say" "that we can speak of a reddish blue but not of a reddish green etc" (PR IV, 39). At first one might find plausible the idea of a geometrical figure "telling" us something by allowing us to see things from a greater distance (Baker 2002: 72). A "geometrical representation" may well suit the task of conveying what by association with (the term) "system" is that without which our terms would not be connected in the special way they are. So far it is the idea of a colour solid's realizing a *structure* in which the trust in its serving its particular purpose lies. This in itself is pronounced to involve "possibilities of colour combinations":

> [o]ne can see in the colour octahedron that some combinations of colour are not possible ('greenish red' while others are allowed for instance yellowish red'). The colour octahedron shows in a glance [...] the different possibilities of colour combination. (Engelmann 2011: 84)

When having a "peculiar face" is described in such a way – what is in question is not only how being a geometrical relation is translatable into the "ish-quality" of colour. Is it that a colour as such limits the range of what can be thought to be "next" to it? Or is it that our concepts give definite content to being the *next colour*? What lies at the bottom of worries concerning this picture of a mutual dependency between colours and concepts has to do with the modality applied here. How do we have to understand "can" in "we can speak of a reddish blue but not of a reddish green" (PR IV, 39; see also what Wittgenstein says about the use of "can" here in his conversation with the Vienna Circle)

Either way: The answer to the question "*Is the colour octahedron a perspicuous representation of the grammar of colour words because the spatial relations between colours correspond to objective/physical features of colour?*" might be yes. Or it might not be, because what is assumed is no more than that the octahedron exhibits "our" grammar by revealing which pairs of colour-terms are interchangeable, and which are not. If it is believed that some colour terms are alike – as "bluish red" and "bluish green" – in respect to being "allowed" and not "prohibited", depending on whether it is *possible* for two colours to join into a relationship, then a certain requirement has to be met. The rules the colour octahedron is said to represent have to be spelled out. Something needs to be said about *colour* as a concept that limits the "possibilities of colour combinations" such as "bluish yellow" or "reddish green".

To say this, however, stands in conflict with what both Schlick and Waismann reported as a result of their discussions with Wittgenstein in 1929 on the question "Would it be possible to discover a *new* colour?" (PR IX, 95)

When Schlick writes in his essay "Is There a Factual *a Priori*?" from 1930:

> [t]he [...] rules which underlie our employment of colour treat [...] the question whether there could be a novel colour as depending on the system of our colour words. (Schlick 1930: 167)

this is meant to put an end to speculations concerning the possibility of colour combinations some fictitious alien grammatical system would ascribe to objects. There would be no way to show the idea of such deviant colour combinations rightfully to have to be excluded, as it is neither the case that

> Red and green are incompatible, [...] because I happen to have never observed such a joint appearance [...]. (Schlick 1930: 169)

nor the case that the assumption of "such a joint appearance" could be proven to involve a *"contradiction in meaning"*. Schlick et al. do not draw the consequence that such a deviant colour system is to be regarded as a genuine possibility. Quite to the contrary: If predicates of this type bear *no* connections to the colour terms we use (they stand in no detectable opposition/contradiction to the meanings of "our" terms), then nothing, no "material", has been made "available" which could be judged in one or the other way.

Having no content, the exclusion of "reddish green" as a possible predicate cannot have its source then in alleged "laws of our thinking" either. Husserl made such a move – wrongly, according to the philosophers of the Vienna Circle. He thought that accepting the negation of the sentence "an object cannot be red and green in the same respect" is not possible for reasons that have to do with the conditions of our being able to *apply* concepts. This at the same time would prove that to entertain the idea of such a deviant concept/predicate involves a contradiction, even if the expression of this contradiction is not the negation of an analytic truth.

For Waismann all that can be said is:

> We do not use words in that way, what you say means nothing. (Waismann 1965: 58)

– which is *his* reaction to the fact that if what there is, is *the only way* things could possibly be, using the category *possibility* is a misnomer. This is due to the struggle of the Viennese with the idea of there being rules that *"underlie our employment of colour"* about which not much more than *this* can be said. Wittgenstein's use of the expression "convention" in the *Philosophical Remarks* is to be understood along the same lines:

> [...] if certain combinations of words had to be excluded as nonsensical, [...] then I cannot cite a property of colours that makes the conventions necessary. (PR I, 4)

> Grammatical conventions cannot be justified by describing what is represented: any such description presupposed the grammatical rules. (PR I, 7)

Nothing then could be *identified* via a structure, if "structure" equals a list of sentences which at best are imagined as being uttered in reaction to a comparison that cannot be pursued. This has consequences even for the "modest" idea that the colour octahedron would be an *"expression* of the rules for the use of colour words" (Hacker 2007) not a list of sentences in which "could" or "could not" occur. But does saying "no Reddish Green", "Reddish Blue – yes" etc. count as contribution to our understanding of colour concepts? Or is a supplementation with "There is …" required so the rules *expressed* yield the sentences "There is no Reddish Green", "There is Reddish Blue" which, in turn, might be thought to be nonsensical – because ultimately metaphysical?

Eventually Waismann abandoned the use of the term "logical structure". Schlick continued to use the term while nonetheless insisting that no *definition* of a colour could be given.

So the whole point of the discussions between Schlick, Waismann and Wittgenstein around 1930 is the question of how to understand *colour*, if *colour* is to be understood as a relationship between a system of colour and those particular colours. The lesson was/is: there could not be any structure, if what is meant by that is something independent of the *colours* the structure is believed to be a structure of.

*

Wittgenstein's treatment of the question about the relation amongst colours in his *Remarks on Colours* is meant to bring that out – to bring out the idea that what is thought of as a geometry of colour is really an illusion.

It might already be obvious why. The aim to exhibit the concept of colour as a system of geometrical relations right from the beginning invites us to interpret "A is more similar to B" as a question about another colour's – C's – relationship. This leads to the question "What is the colour x a colour y is similar to?" – for example "what is the colour which is similar to yellow?". In order to solve this task one has to answer the further question "What is placed next to yellow in such a way that the colour that follows next is of the same relationship to the colours it is in-between, as the one it itself is followed by?" And so on. The idea here is one of a mutually mediated (con)sequence, or, in other words, that the "between-ness" relation could account for similarity and contrast among all colours.

But one only has to acknowledge what is used here as a "method" of colour determination in order to see that inevitably differentiation(s) in colour are not accountable any longer. The definite answer to the question of what is yellow's neighbour is allegedly: *Green*, not *Blue!* The reason given is that this sequence alone – the triple yellow-green-blue – assures us that the colour that is next to what is next to yellow can be positioned *between* two colours in a way that equals yellow's neighbour (which in blue's case would yield standing in-between green and red). The result reached by applying this procedure is remarkable, yet not surprising: *all* the primary colours, i.e. yellow, blue, green, red are alike. If yellow, green, blue, and red do not stand in a "between-ness" relation peculiar to *them*, they are alike to what is also *between* some other positions – yellowish green, turquoise, purple, orange. No "Gegensatz und Verwandtschaft" in colour so far (RC III, 46).

One might understand "Blue is more similar to purple than to yellow" (the example in the PR passim, see also RC I, 50) such that blue is different from green by standing in the "between-ness" relation to two *different* colours. But if being *blue* is the same as being *between* those two colours, then one has not given any independent specification of what blue is. Therefore either some understanding is presupposed or one does not really know what blue is. In the latter case one might then ask: "In virtue of what does blue stand in the between-ness relation to those two colours?"

The colour octahedron is no answer to this question. This is so, not just because as a picture it is not supposed to be one. It makes no sense to talk about one's colour being more similar to another colour, if all differentiation(s) in colour is annihilated. Wittgenstein *explicitly* says this with respect to the "double-cone" (PR XXI, 221).

But if it really all depended on colours' being put in how many angles of 90 degrees, where "angle of 90 degrees" must not be understood in a geometrical sense at all, but only as "metaphor" ("Gleichnis") (PR XXI, 221), then the charge applies all the same: A geometrical representation of colour relations fails on its own account.

*

What Wittgenstein failed to acknowledge in his own words was:

> [...] that we have not *one* but several related concepts of the sameness of colours. (RC III, 251)

Later on, after his "phenomenological period", he will suggest that although the colour octahedron was to be a model of grammar distinct from the one he once defended, it proves to be the 'worthy' heir of the calculus view of grammar.

Wittgenstein's interest in sentences expressing relations among colours orginiated from recognizing that the interpretation of "a is blue and red" along the lines of "a is blue and oblong" yields a result standing in contrast to how we comprehend such a sentence. So this interpretation is mistaken: "a is blue-red" does not represent "a logical product of 'a is blue' and 'a is red'"(PR VIII, 80). Adhering to the TLP account of meaning "a is blue-red" points towards an investigation of what could be shown about the negation of (colour-) predicates in *general* ("as a system"). It is in relation to this aim that there being differences in the combinations of different colours becomes of interest for Wittgenstein.

> One's first thought is that it is incompatible for two colours to be in *one* place at the same time. The next is that two colours in one place simply combine to make another. But third comes the objection: how about the complementary colours? (PR VIII, 76)

But if it is true that

> one measurement [...] automatically excludes all others (PR VIII, 80)

> [...] a rod can only have one length, a man only one age and a point in space only one temperature. Exactly the same is true of "colour" (Waismann 1965: 59),

these differences do not matter. The reason, however, is not, and was not for Wittgenstein at that time either, that particular colours were degrees of colour. It is simply this: what someone meant to say by for example "x is red *and* pink" is difficult to figure out. In this respect there is no difference from the puzzlement one would feel, if someone called the colour of an object "red *and* green". If one only had these sentences in order to understand, one would not know what to think – in both cases.

When Wittgenstein presents later – in the *Remarks on Colour* – a language game in which "-ish expressions" are to be learned as a situation where things are to be "ordered" or "selected",

> How do I learn use the word "yellowish"? Through language-games in which, for example, things are put in a certain order (RC III, 110)

this is to signify a contrast. What one is *not* supposed to be taught is that pink objects are so *by* being red – be it by being of a certain degree or any other way of indicating their being in the "extension" of *red*. Saying "A is red and pink" is to tell the pupil that among the red objects there are some objects, which *if* they have *this* colour are called "pink". There is then a difference between "A is red and pink" and "Aristotle is a philosopher and wise". The "and" in "x is red *and* pink" is no version of "*if* philosopher *then* wise". Since it does not accord with our understanding of "red" or "pink" that objects are pink by being red, "being pink" is not subsumable under "being red" in the way "being wise" is subsumable under "being a philosopher".

Wittgenstein fully acknowledged that being of a certain colour cannot be a criterion of excluding another colour in the way which "being a philosopher" excludes "being not wise". He would not have felt that he had to investigate the inter-relationships of colours in the peculiar way he did, had he thought he could appeal to this logic-textbook view of concept subordination. To seek a conceptual analysis of colour that does justice to this fact and proceeds independently of the usage of colour concepts in predication, however, leaves one with exactly the "and" view of colour-*ascription* that was identified as a source of "difficulties" in understanding sentences using colour concepts: A "redder-blue" cannot be accounted for with "the 'and' of logic", since it does not make any sense "to say that a rod which is 3 yards long is 2 yards long, because it is 2 + 1 yards long [...]" (PR VIII, 76).

That Wittgenstein thought that where the TLP had failed the colour octahedron could succeed, so that which negations (not pink, not blue, not red) exclude which colour could be shown in a systematic way is an inconsistency. This, however, should not make one appreciate less what he says about the concept *colour* in the TLP.

*

Wittgenstein's fundamental insight is that *colours* are not used to describe *how* things are. In the TLP this is expressed by putting *colour* in the neighbourhood of concepts where to say that some (spatial) objects can be subsumed under this concept makes no sense given that "(a) speck in the visual field [...] *must* have some colour. [...]" (TLP 2.0131, my italics, G.M.M.).

> Space, time, colour (being coloured) are forms of objects. (TLP 2.0251) [...] the pseudo-concept object.

> [...] the word "object" ("thing", etc.) is correctly used [...] in the proposition, "There are 2 objects which [...]", [...]. Wherever it is used in a different way, that is as a proper concept-word, non-sensical pseudo-propositions are the result. So one cannot say, for example, "There are objects", as one might say, "There are books". (TLP 4.1272)

and again in *On Certainty*:

> "A is a physical object" is a piece of instruction which we give only to someone who doesn't yet understand either what "A" means or what "physical object" means. [...] "physical object" is a logical concept" (Like colour, quantity, ...) [...] (OC §36).

Early or late, Wittgenstein wanted to repudiate the idea that concepts of colour are to be understood on the pattern of the analogy of class membership. There is nothing in addition to being blue that distinguishes blue from green. Why does this matter? It does so because then there is *no open* question for us, something to inquire into as far as the relation between colours is concerned. The appropriate reaction to "How does being blue differ from being green?" is simply: "by being *blue*". Colours do not differ from each other in any other way except by being *not* what any another colour *is*.

What may make this difficult to acknowledge is that:

> "The colours" are not things that have any properties whatever [...]. (RC III, 127)[1]

i.e. that colours do not themselves have any properties beyond being the very colours they are, namely being blue, yellow, green and so on, seems to leave us without any means to account for what is a (vital) part in seeing, looking for, choosing colours.

If one thinks, as Wittgenstein did in the *Remarks on Colour*, that any attempt to understand colours which invokes a property instead of the property *colour* is doomed to failure, then it sounds as if *no* questions about combination of two particular colours could be raised. Yet the idea of such combinations seems to make sense because we understand "one colour is more similar to another than to a third" as allowing us to use expressions like "bluish red".

One could understand what Wittgenstein does in the *Remarks on Colour* as being due to this "tension". He presents various unities – in which colours are put together "on the same level" as in a flag or in a Kodak colour picture of the Austrian alps (RC III, 1,2, cf. RFM 42, RC III, 13,17). This allows for colour sequences different from the "standard" colour circle.

> In a brightly coloured pattern black and white can be next to red and green, etc. without standing out as different. This would not be the case, however, in the colour circle, [...]. (RC II, 85)[2]

That there will always be some respect in which colours are arranged as more or less 'similar' so that their order falls outside the frame of the colour solid might invite the thought that the idea of an alien colour grammar is not inconceivable. Nonetheless, for Wittgenstein it is not "without any further ado" possible, "to begin to search for or conceive of colours we don't know yet [...]." (RC III, 127)[3]. He continues to use phrases which suggest that some colours "fit" others best and some others do not fit together

---

[1] Orig.: "'Die Farben', das sind nicht Dinge, die bestimmte Eigenschaften haben, [...]". I deviate here from translating 'bestimmte' as 'definite'. The German use of the expression 'bestimmte' is in this context equivalent to 'irgendwelche', i.e. saying 'whatever properties', 'it does not matter which'. The translation in Wittgenstein 1977 is 'The colours' are not things that have definite properties ...'; see also footnote 2.

[2] Orig: "In einem bunten Muster könnte Schwarzes und Weißes neben Rotem und Grünem etc. sein, ohne als andersartig sich abzusondern. Nur im Farbenkreis fiele es heraus." (RC II, 85, see also 2)

[3] The translation here deviates again from the standard translation. The way "ohne weiteres" is translated does not make it sufficiently clear that if we had to search for colours "unknown to us", we would have to have some criteria, i.e. knowledge of some properties of what we are looking for, in order to begin the search at all. But as Wittgenstein writes, colours do not have what is required here: "*'Die Farben', das sind nicht Dinge, die bestimmte Eigenschaften haben, [...]*".

at all (as "transparent white" or "reddish green", see also PR IX, 95). So the question "What does it mean to say, 'Brown is akin to yellow'?"[4] is not the expression of any doubt concerning the intelligibility of calling brown and yellow "verwandt in Farbe" as Wittgenstein still writes: "In den Farben: Verwandschaft, und Gegensatz. (Und das ist Logik.)" (RC III, 46) "Among the colours: Kinship and Contrast. (And that is logic.)"

Do we really have any reason to believe that the late Wittgenstein thought that the tension involved in both these claims could be overcome? – By the idea of a (single) colour-pair that *independently* of any context would enable us to look upon all *other* colour-pairs as derived from this one couple and distinct from it merely as a matter of degree? If so, we would be back at the idea of the colour octahedron.

So the response to this attempted reconciliation of "internal" and "external" ought to be abandoned – as any search for something that could represent an answer to the question "In *virtue* of what is blue more similar to purple than to yellow?" It is equally necessary to understand that the *respect* in relation to which something is said to be similar does not coincide with *what* it is that is thereby said to be similar. In order to account for the meaning of "*more similar*" it is *futile* to look for some thing/aspect with which this relation could be identified. This brings me back to the use of colour concepts in predication.

What was said to be at stake is the extent to which our concept of colour allows an understanding that rests on grounds independently of any use of colour concepts as applied in predication. This concern has to be distinguished from the question whether using colour terms is exhausted by predicating colour terms to objects – something that is certainly not true and not Wittgenstein's view. But it is Wittgenstein's view that the use of colour concepts rests on a distinction that could not be accounted for in the colour octahedron. This does not mean one could not distinguish the usage of colour concepts in predication from their being used to say something about these concepts as "objects". Since no criterion of identity of colour *concepts* can be given – a colour (property) cannot *by* being a particular property be *the same* as a colour property distinct from it. *Objects* can be said to be the similar in colour. So it seems that in relation to objects the ascription of one colour property can be understood as linked to another colour property.

It is in using "…ish" expressions, "bluish green" etc., that we show our understanding of this linkage. It is also in connection with "-ish" expressions that Wittgenstein in the *Remarks on Colour* uses "internal relation", i.e. speaks of what has to do with "the logic of colour". Accordingly some people thought of having a "different geometry of colour" as one's having "a different concept of ' …ish'" (RC I, 30). In the manuscripts published as *Remarks on Colour* and as *On Certainty* Wittgenstein treats questions concerning the understanding of "…ish" as being connected with the concept of *shades* of colour ("Farbton"). Here

---

[4] Was heißt es, 'Das Braun ist dem Gelb verwandt'? (RC III, 47)

> [...] the relationship between the lightness of certain shades of colour. [...] is an internal relation [...]. (RC I, 1)

Thus a language game as imagined above – "object a is red and pink" – is peculiar insofar as saying something true of an object also involves some respect in which the colour property is presented. The concept of *shade* here (RC I, 34) is to be distinguished from that of the concept of a shade in virtue of which all colour are made alike as degrees of brightness ("Colours are already a shadow" LWPP I, 214; see too: RPP II, 398, RC III, 57). This would be the very understanding that denies significance to the "internal relation" said to be expressed in "being more similar to" or "bluish red".

> A less yellow green [...] is not a bluer one [...] (RC III, 158)

> [...] saturated [...] Yellow [...] is lighter than red. Is red lighter than blue? I don't know. (RC III, 161)

> Why is it that a dark yellow does not have to be perceived as "blackish", even if we call it dark? (RC III, 106)

There is an attempt to respect this connection in the *Remarks on Colour*. The view Wittgenstein considers is the view of Brentano, as Schulte has drawn our attention to in his "Mischfarben. Betrachtungen zu einer These Brentanos und einem Gedanken Wittgensteins" (Schulte 1990). That view opposes the whole idea of one colour as being between two other colours, but there is in itself no difficulty in one colour being between two other colours in this way. What Brentano really opposes is the idea that the betweenness relation holding between two colours can account for our understanding of colour terms. He appeals instead to the idea of mixed colours, and he seems to conclude that when he sees what he thinks of as a mixed colour like green he actually sees it as a mixture of blue and yellow.

What Wittgenstein appears to be saying in 26 and 40 in the *Remarks on Colour* III is that the application of colour predicates to perceived objects is not done on the basis of perceiving the colours as a mixture of two other colours – there is a discordance between what the alleged geometry of colours would say in terms of mixtures and what we see and think of being true of objects when we predicate colour terms to them.

The problem is that the so called geometry of colours expressed in any form does not seem to explain the character of the colour predicates we know we apply to objects when we say the objects are of this and that colour. The trouble with the colour geometry is that it seeks to identify something that colours have in common. But what any colour has in common with all other colours cannot account for the distinctive character of that particular colour. To acknowledge this while retaining the idea of a geometry of colours would mean that the distinctive character of a particular colour could be identified only by its particular position in its geometrical space.

There is no alternative to this idea of a geometry of colours, if one thinks it as a way whereby one ultimately could account for what it is that makes colours distinct and thereby allows for *some* combinations and not for others. Since colours are not the same in virtue of being colours their being distinct cannot simply be identified by any property. Wittgenstein's conclusion:

> What are we to say are propositions about blue? We could say two quite different things. We might say "There are blue books here" is a proposition about blue; or, thinking that it is not about blue because "blue" is only an adjective there, we might say "blue is darker than yellow" is a proposition about blue. I say the way I'll go is the first way. (LFM 250)

References

Quine, W. V. O.: *Pursuit of Truth*. Second edition. Cambridge, Mass.: Harvard University Press, 1992.

Gary Kemp
Did Wittgenstein have a Theory of Colour?

Philosophers are often concerned to describe the nature of things, and in particular the nature of colour. I've often been puzzled as to why it is assumed that there is a nature of colour to be described, as to why anyone would think that there is. So with the assistance of Wittgenstein, I will try to articulate why I think that there isn't such a thing.

Thus I am concerned with a somewhat different question about Wittgenstein's view or thinking about colour from those explored in the other essays in this volume. In particular, I will not add anything substantial to the various remarks and interpretations offered there of the *Remarks on Colour*. Actually my question presupposes or suggests much that Wittgenstein would have no part of: for I'm sure that Wittgenstein did not simply lack such a theory in the way that he lacked a theory of fly-fishing, in that he was unable or unwilling to formulate such a theory. My sense is that he thought that, although of course there are many true scientific propositions about colour, the idea of a *philosophical theory* of colour is just confused; there is no such thing as a true philosophical and informative proposition about the nature of colour, and indeed the whole idea doesn't really make sense. In a way that ought to be obvious to anyone who sympathises with such Wittgensteinian sayings as that 'if someone were to advance theses in philosophy, it would never be possible to debate them, because everyone would agree to them' (PI, §128); '...we may not advance any kind of theory' (PI, §109). For what it's worth, this anti-theoretical attitude is one that I share, at least on the present subject. And it is especially interesting to see how this attitude plays out in the case of colour, for it can seem quite wrong to suppose that there is no fact of the matter with respect to certain philosophical questions about colour.

Wittgenstein did suppose that there exists a 'grammar of colour', a 'language-game of colour'; these comprise such statements as 'red is darker than pink', 'the complement of blue is orange', and so on. These can be arranged to show their various relationships – as for example in the colour wheel – which constitute a 'logic of colour', in Wittgenstein's way of talking. These do have a special status; they seem to be conceptual truths, even if that must be understood with an enormous pinch of salt in the context of Wittgenstein exegesis. They are not analytic truths – unless the notion of analyticity is widened from a Fregean understanding of it as logical truths plus statements that can be converted into logical truths by substituting synonyms. They are rather truths such that acceptance of such a statement is a criterion for understanding the statement – or rather, more broadly, for speaking as we do, for taking part in our ordinary language-game of colour. Happily nothing I will say depends on how precisely Wittgenstein understands 'understanding', or on a precise exegesis of the notion of grammatical propositions. For however these are explained, they do not determine answers to the philosophical questions about colour I have in mind.

If I had to say what what was my Wittgensteinian watchword, it would be – even though it was directed towards another subject – §79 of *Philosophical Investigations*: 'say what you please, so long as it does not prevent you from seeing how things are'.

1 Shaping the Question Further

In asking about 'theories of colour' I mean the attempt to answer the philosophical question "What is colour?"; it is the attempt to elucidate the *nature* of colour, its *essence*. I'm going to sharpen the question in terms of a contemporary framework of possible worlds. This might be thought misleading in a paper primarily about Wittgenstein, who would presumably have taken a dim view of such talk. But for my purposes we have to allow the question to be formulated with some degree of precision, to give it some traction. I shall assume, however counterfactually, that Wittgenstein would not object to the very formulation of the issue. Thus if the answer to the question about colour is X, then that colour=X is (at least) a necessary truth, a proposition that is true in all possible worlds. This is familiar in recent times from the work of Saul Kripke and Hilary Putnam.

There are nowadays many, many competing accounts of the 'nature of colour'. Most of them correctly predict our actual use of colour-terms in everyday, normal circumstances. They diverge in what they say about counterfactual circumstances, what they take to be possible, and so on. My view, roughly, is that any view that gets ordinary statements about the actual world right is as good as any other, and that beyond that rather low hurdle, there is no fact of the matter concerning which is right. Unfortunately I'm not going to *argue for* the view, because I doubt whether I could; I should be glad if I succeeded in explaining something of why the view is true if it is true.' Perhaps it will be said that my task is therefore, at best, one of persuasion, of rhetoric. But I like to think that the view is the view of good sense, and I'll try to make it plausible that it was Wittgenstein's view (for hints of my subject: RC III, 251-265).

There are broadly speaking three sorts of answers to the question: (1) The fully objective, scientific essentialist answer; according to which the nature of colour is like that of magnetism, revealed by scientific investigation. (2) The phenomenalist or phenomenological answer, according to which the nature of colour is found out by introspection, or reflection on the experience of sensing colour, and the like. (3) The linguistic or conceptual answer. In Wittgenstein's terms, of course, it is where one speaks of the logic or grammar of colour; the nature of colour is decided by answering such questions as: How is the word 'red' used? 'What is the criterion of sameness of colour?' – where this is determined by linguistic practice, by language-games.

I will not consider the possibility that these questions are more or less separate, or that the answer to one of these tells you the answer to the other, or that they interact more subtly. Another possibility is that they're not in competition, that they're concerned with different 'concepts' of colour; this possibility I'll return to briefly at the end.

My answer is going to be that Wittgenstein accepts neither scientific essentialism about the nature of colour, nor a phenomenalist account – but that neither does he accept that, in the sense relevant to my question, linguistic practice or language-games reveal the essence of colour or colours. It's true that 'Essence is expressed by grammar' was an important remark of Wittgenstein's. Maybe insofar as there are essences, they are according to Wittgenstein reflected or determined by grammar; but really there aren't any such things, at least not of the sort that would satisfy metaphysicians or philosophers of science today. I think when Wittgenstein said essence is expressed by grammar, he meant only that grammar reveals all there is to essence, not that the end of a grammatical investigation will reveal entities that retain their identities across counterfactual situations or possible worlds; on the contrary, I think that the vocabulary of colour falls apart when subject to such rigour. There are obvious truths about colour, and there are things that upon careful reflection we can see are true, but these are all there are about colour, and they fall far short of the essence of colour in the metaphysical sense.

2 Some Answers Rejected

I'll sketch an answer to the question about the nature of colour which I believe to be correct, before turning explicitly to Wittgenstein.

Thus let us look a little more closely into the first two views of colour. First, again, scientific essentialist views. Some are impressed with Kripke and Putnam's apparent derivation of the essence of water; and they will say that what it is to be red is necessarily to have a certain reflectance property (a disposition to reflect a certain wavelength of light; really I just use 'reflectance property' as a stand-in for any property that physics might single out as the colour-property of an object, what used to be called the primary qualities underlying colour). This is not to deny that if they had had different reflectance properties, it is still possible for things such as ripe strawberries to in some sense to have appeared to us as they actually appear. For we can embrace semantic Two-Dimensionalism, according which there is Metaphysical Possibility – which Kripke taught us about – and there is also If-this-had-been Actual-Possibility. Suppose that colour tracks reflectance properties across possible worlds. Yet we seem to be able to describe a world with different reflectance properties but at which we – perhaps because we have different lenses in our eyes – still have the same chromatic sensation when looking at a ripe strawberry as we have in the actual world. Such a world shows that different reflectance properties *might have been* denoted by the words, that the stereotypes associated with words might have determined different reflectance properties, different colours. According to the scientific essentialist outlook on colour, then, it is open to say that this does show possibility, though only in the second type of possibility.

Second, phenomenal views. Others will say that colour is not like that; 'red', for example, is not a rigid designator standing for a reflectance property. If R is the chromatic sensation actually caused by Red things, then 'Red' tracks across worlds the dispositions of objects to cause R. So it is with the idea of sweetness – a non-rigid, response-dependent but intersubjective property of objects. Redness, like sweetness, is multiply realisable; in particular it is multiply realisable with respect to reflectance properties. All that matters is how the strawberry looks and tastes.

Let us call the chromatic sensation we actually have when looking at red things or substances the R-sensation, and call the things and substances that are actually red, the R-stuff (ignore the complications induced by phenomenon of iridescence, and the colours of amorphous things like gas clouds or the sky). Likewise for G-sensations, G-stuff, and the colour green. Necessarily, something is an instance of R-stuff if and only if a fundamental molecular description applies to it that applies to actual R-stuff; similarly for G-stuff.[1]

In possible-world W1 there is a systematic change from the actual world W@ in the gasses making up the atmosphere on such planets as Earth, so that R-stuff does not give rise there to the R-sensation, but gives rise instead to a G-sensation (in language of the actual world, it 'looks green'). It is not a difference in reflectance properties, but the gasses act as an inverting filter on the light reflected. Is the R-stuff – using the words with the same meanings as we do in the actual world – red, or green, or neither?

Now consider W2. It differs from W@ only in our retinas (or optic nerves, or anything so long as it is inside the bodily envelope): Looking at R-stuff, they send the signals downstream that in the actual world their analogues send when looking at G-stuff. The R-stuff thus retains its actual reflectance properties and effects in atmosphere. Yet still, there is strong temptation to say that in the actual world language, R-stuff 'looks green' (again, speaking English). But is it red, green, or neither?

I find I am pulled in the 'subjective' phenomenal direction by that thought that what normally appears green is green; that is, if it is normal in W1 and W2 to have R-stuff giving rise to a chromatic sensation which in the actual world is had when looking

---

[1] Thanks to Barry Smith for drawing the point out. By a 'fundamental molecular description' I mean, for example, a description of water as comprising two hydrogen atoms covalently bound to one oxygen atom. Needless to say, a fundamental molecular description of all the R-stuff would be enormously complex. As an alternative partial characterisation, we could say that an object in a counterfactual situation is an example of R-stuff if it is identical at the molecular level to an actual example of R-stuff. The idea is that if you take an actually red object and consider it in alternative world, it will remain R-stuff, however it appears chromatically.

at G-stuff, then R-stuff, in W1 and W2, is green. But I am pulled the 'objective' scientific direction by the fact nothing in the objects, and nothing in the environment of W2, has changed; surely no fact about ourselves can bear on the colours of things. But if we're taking this line, we cannot fail to be impressed by a certain affinity of W2 with W1: If R-stuff is green in W1-circumstances, then surely it is in W2; hence by *modus tollens* things that are not green in W2 – they remain red – are not green in W1.

Of course my view is that there is no fact of the matter; moreover, I cannot for the life of me see why anyone supposes there is. Our colour vocabulary is made possible by many interlocking contingencies. 'Contingency' makes it sound as if it might at any second all collapse into nonsense, but that is not so. There are of course *some* 'facts of the matter' about colour; describe all the facts that I gestured at above, and describe the facts underlying our use of colour-words, and you have described those facts: Certain substances or materials, their molecular configurations; the reflectance patterns, wavelengths of light travelling through space; the actions our lenses; the effect on our retinas; the amazingly intricate neurological processes in the optic nerves, the complicated events that take place in the brain; the 'sensation' or 'experiences' caused, the patterns of speech that form. A certain constellation of such facts, a whole sequence of circumstances, is involved. These arrangements and mechanisms evolved over a very long time, and now work with a very high degree of reliability. In the actual world, that sequence is rarely disturbed in a way that does not simply leave the person blind or colour-blind; certainly nothing like the scenarios described under W1 or W2 ever actually are encountered. So why think that even so, there must be an answer in respect to W1 or W2 (or any of the other variants on that theme)? Nothing, I suspect, except the curiosities and presumptions, and sometimes the obsessions, of (some) philosophers.

My claim, to repeat, is not there is no fact of the matter about colour. It is that there are no facts corresponding to (many of the) modal questions about colour that metaphysicians like to pose. In that sense, colour has no essence.

Before turning back to Wittgenstein, I should consider a certain response on the part of certain metaphysicians: 'Aha! You've gestured at a list of what you call actual facts of the matter concerning redness, but unless you say which are counterfactual supporting, which are necessary and sufficient for the truth of colour-judgements, you haven't specified the *property of redness*, and thus your claim that there is a fact of the matter about redness is unfounded. In fact, all properties remain the same in counterfactual circumstances, just as objects remain numerically the same if they exist at all. So you cannot say there is a fact of the matter about a property without saying that there are modal facts about it.'

To this I will say, conforming to the view of properties being assumed by the metaphysician, that colours, in the ordinary senses of the terms, are not properties. Properties are what we can speak confidently of across counterfactual situations; mostly these will be the properties that figure in physics and chemistry, such as *conducts electricity* or *is a noble gas*. A predicate like 'is red', could be described as standing for a *configuration* of properties (the ones figuring in my list), but this does not mean that it does so in a counterfactual-supporting way: in counterfactual situations where the actual configuration of facts is not realized, the word's actual meaning does not – or does not always or as frequently as metaphysicians would like – determine its extension. To be more precise: Call a world in which the configuration of chromatic-relevant facts is crucially unlike the actual configuration a 'chromatically challenged' world. (I would say also that the line between the chromatically challenged worlds and the rest is indeterminate or vague.) Then we can put my thesis as this: The Kaplanian *character* of 'red' is such that it falls apart, lapses, fall silent, with respect to chromatically challenged worlds. Linguistic norms arise in the actual world, and have implications for alternative counterfactual situations only in certain dimensions.

3 Support from Wittgenstein

So why do I think Wittgenstein would agree? (Or, not to be too presumptuous: why do I think I am following Wittgenstein in taking such a view of the problem?)

One so easily thinks that one's grasp of concepts – one's thinking or understanding – is impermeable in a way that ordinary knowledge of matters of fact is not. We have an almost indefatigable allegiance or temptation to believe in hyper-certainty with respect to the conceptual realm, or to thought, or to the contents of one's mind – for something that remains the case whatever happens in the world. But there simply is no sublime realm of super-hard facts or rails to infinity; Wittgenstein at PI §80 (see also BB, p. 27):

> I say, "There is a chair over there". What if I go to fetch it, and it suddenly disappears from sight? – "So it wasn't a chair, but some kind of illusion." – But a few seconds later, we see it again and are able to touch it, and so on. – "So the chair was there after all, and its disappearance was some kind of illusion." – But suppose that after a time it disappears again a or seems to disappear. What are we to say now? Have you rules ready for such cases a rules saying whether such a thing is still to be called a "chair"? But do we miss them when we use the word "chair"? And are we to say that we do not really attach any meaning to this word, because we are not equipped with rules for every possible application of it?

A sufficiently large disturbance in the world seems to infect the stability or determinacy of meaning, of concepts. One thought one knew what chairs were, one thought one's grasp of chairhood was secure, but in describable circumstances one finds that concept slipping through one's fingers. Of course what is described is not something that ever happens. The point is that concepts need only be as sharp as the world demands in practice. 'It is idle to brook definitions against implausible contingencies', wrote Quine (1992 p. 21); likewise at PI §87 Wittgenstein says 'The signpost is in order – if, under *normal circumstances*, it fulfils its purpose.' (emphasis added; see also PI, §85).

The same point can be extracted from another famous passage from the *Philosophical Investigations*; I quote it at some length:

> If one says "Moses did not exist", this may mean various things. It may mean: the Israelites did not have a single leader when they came out from Egypt – or: their leader was not called Moses – or: there wasn't anyone who accomplished all that the Bible relates of Moses – or: ... – According to Russell, we may say: the name "Moses" can be defined by |37| means of various descriptions. For example, as "the man who led the Israelites through the wilderness", "the man who lived at that time and place and was then called 'Moses'", "the man who as a child was taken out of the Nile by Pharaoh's daughter", and so on. And according as we accept one definition or another, the sentence "Moses did exist" acquired a different sense, and so does every other sentence about Moses. – And if we are told "N did not exist", we do ask: "What do you mean? Do you want to say ... or ... and so on?"
> But if I make a statement about Moses, am I always ready to substitute some *one* of these descriptions for "Moses"? I shall perhaps say: By "Moses" I mean the man who did what the Bible relates of Moses, or at any rate much of it. But how much? Have I decided how much must turn out to be false for give up my proposition as false? So is my use of the name "Moses" fixed and determined for all possible cases? – Isn't it like this, that I have, so to speak, a whole series of props in readiness, and am ready to lean on one if another should be taken from under me, and vice versa? – Consider yet another case. If I say "N is dead", then something like the following may hold for the meaning of the name "N": I believe that a human being has lived, whom (1) I have seen in such-and-such places, who (2) looked like this (pictures), (3) has done such-and-such things, and (4) bore the name "N" in civic life. – Asked what I mean by "N", I'd enumerate all or some of these points, and different ones on different occasions. So my definition of "N" would perhaps be "the mean of whom all this is true". – But if some point were now to turn out to be false? – Would I be prepared to declare the proposition "N is dead" false – even if what has turned out to be false is only something which strikes me as insignificant? But where are the boundaries of what is insignificant? – If I had given an explanation of the name in such a case, I'd now be ready to alter it.
> And this can be expressed as follows: I use the name "N" without a *fixed* meaning. (But that impairs its use as little as the use of a table is impaired by the fact that it stand on four legs instead of three and so sometimes wobbled.)
> Should it be said that I'm using a word whose meaning I don't know, and so am talking nonsense? – Say what you please, so long as it does not prevent you from seeing how things are. (And when you see that, there will be some things that you won't say.)
> (The fluctuation of scientific definitions: what today counts as an |38| observed concomitant of phenomenon A will tomorrow be used to define "A"). (PI, §79)

Several interwoven themes emerge from this. One idea that has been extracted from it is that of a 'Cluster Theory' of proper names: Perhaps, rather than a name's being synonymous with a simple definite description, it is to be equated with a more complicated description, something like 'the object of which most of the following are true', followed by a list of predicates, perhaps weighted in such a way that some combinations of them are more important than others for identifying the object. Alternatively, the cluster idea might be simply that the name is synonymous with a disjunction of conjunctions: N is the man who did either A, B and C, or A, B and D, or B, C and D, or .... But we need not stop over details, as this is definitely not the sort of thing that

Wittgenstein is suggesting. Wittgenstein is denying that there is any single rule governing the use of proper names: if we look carefully at how a (reasonable) speaker uses proper names, and the sorts of things he would say in response to various possibilities, the idea that there must be such a rule, or the idea that one is needed in order to characterise the correct use of proper names, evaporates. Again, Wittgenstein's lesson is that in *sufficiently unusual* circumstances, language, or meaning, falls apart, gives no hint as to how to go on. But since these are highly anomalous circumstances – the sort of fanciful things dreamed up by philosophers – it's no wonder that language lacks rules for them. The lesson is repeated in various contexts.

So if we apply that lesson to colour, it seems obvious that my statement that there is no fact of the matter concerning the colour of objects at W1 and W2 is according to Wittgenstein a thing that one ought to affirm: the scenarios are manifestly out of the ordinary, and language is going on holiday. I said above that our colour vocabulary is made possible by many interlocking contingencies, like a Rube Goldberg device. The fact of normally seeing red depends upon imponderably many such facts. If we try to remove or alter them in thought, and ask 'Would an apple still be red?', there needn't be an answer. The smooth operation of colour-language is like a skater dancing, oblivious to the complex objects and forces that make her skating possible. In the *Investigations*, the fact that words require a lot of stage-setting to have the significance they have is stressed throughout, from the beginning language games; only with a background of certain, somewhat sophisticated forms of life can an utterance of 'slab' have its significance.

In other possible worlds, or at least at other conceivable worlds, these contingencies break up. In one, different molecular configurations reflect different wavelengths of light; in another, the atmosphere shifts the reflected wavelengths; in another, the lens behaves differently; in another, the retina behaves differently; in another, events in the optic nerve are from the point of view of the actual world scrambled; in another, the brain responds differently; in another, perhaps, the experience is the only one that is different. Or there are more than two things that are different. Or three, and so on. But I don't see why, speaking of course our language, there has to be an answer as to colours of things in such situations, and I think it clear that Wittgenstein didn't either. There is no problem to which such an answer is addressed.

I said that another possibility is that the three methods appertain to different concepts of colour, or different sharpenings, or different Carnapian explications, of the ordinary concept of colour. I'm not dead set against this, provided that the phenomenological answer really makes sense (again, perhaps the private language argument undermines it; see RC III, 248). But I don't quite advocate this view because I can't see an ambiguity in colour concepts that demands resolution. We can *decide* what to say, but why? Such a move would be empty unless we had some definite purpose in mind. I say that things that don't matter, don't matter.

4 Remark on the Remarks

Of course there are many local, small-scale problems concerning colour which have the appearance of being genuine; some of those form the topics addressed by the *Remarks on Colour*. I leave off with a suggestion about a key topic of the *Remarks*.

It concerns Wittgenstein's interest in a mathematics or logic of colour. I for one think that the sort of attitude just canvassed does have some further, more specialised applications. Compare the grammar of the colour-vocabulary with the Peano axioms for 1st order arithmetic: it is hopeless to settle the standard model, but adequate in practice for solving arithmetical problems. Granted, the grammar of colour is, on the one hand, much richer than a set of axioms for the natural numbers, and on the other I suspect not nearly so determinate, and more variable and subject to outside influences. Can there really not be 'transparent white'? Wittgenstein, in the *Remarks on Colour*, was tortured by this, and for the very good reason that our language-game, the logic of colour concepts, seems here to falter, yet still it seems for all the world a well-formed question. I'm not sure why he thought there must be an answer to this, and why he seems persuaded that transparent white is impossible. One could challenge him with certain examples of white media that are not merely translucent but through which one can see images (mist, for example, or milk watered down). I am open to persuasion, but I don't think there need be an answer – not if one is looking to answer a question couched in our ordinary concepts. Some questions that puzzle us most deeply turn out not to have answers; they turn out in the end like 'What time is it on the North Pole?'.

References

Baker, G. P.: Philosophical Investigation Section 122. Neglected Aspects. In: Shanker, S./ Kilfoyle, D. (Eds.), *Ludwig Wittgenstein. Critical Assessments of Leading Philosophers*, p. 68–94, London: Routledge, 2002.

Engelmann, M. L.: What Wittgenstein's Grammar is *Not*. On Garver, Baker and Hacker, and Hacker on Wittgenstein on 'Grammar'. In: Lütterfelds, W./ Majetschak, S./ Raatzsch, R./ Vossenkuhl, W. (Eds.), *Wittgenstein Studien*, vol. 2, p. 71–102, Berlin, New York: de Gruyter, 2011.

Hacker, P. M. S.: Gordon Baker's Late Interpretation of Wittgenstein. In: Kahane, G./ Kanterian, E./ Kuusela, O. (Eds.), *Wittgenstein and His Interpreters: Essays in Memory of Gordon Baker*, p. 88–122, Oxford: Blackwell, 2007.

Schlick, M.: Is There a Factual *a Priori*? In: Schlick, M./ Mulder, H./ Schlick, M./ van de Velde-Schlick, B. (Eds.), *Moritz Schlick: Philosophical Papers (1925 - 1936)*, vol. 2, p. 161–170, Dordrecht: D. Reidel, 1979.

Schulte, J.: Mischfarben. Betrachtungen zu einer These Brentanos und einem Gedanken Wittgensteins. In: *Chor und Gesetz - Wittgenstein im Kontext*, Frankfurt a. M.: Suhrkamp, 1990.

Waismann, F.: *The Principles of Linguistic Philosophy*. London: Macmillan, 1965.

Frederik A. Gierlinger
"Imagine a Tribe of Colour-Blind People"

> "Imagine a *tribe* of colour-blind people, and there could easily be one. They would not have the same colour concepts as we do. For even assuming they speak, e.g. English, and thus have all the English colour words, they would still use them differently than we do and would *learn* their use differently.
> Or if they have a foreign language, it would be difficult for us to translate their colour words into ours.
> But even if there were also people for whom it was natural to use the expressions 'reddish-green' or 'yellowish-blue' in a consistent manner and who perhaps also exhibit abilities which we lack, we would still not be forced to recognize that they see colours which we do not see. There is, after all, no *commonly* accepted criterion for what is a colour, unless it is one of our colours." (RC I, 13-14)

In this passage from *Remarks on Colour* we are invited to imagine a tribe of colour-blind people. Wittgenstein writes "There could easily be one" and suggests that these people "would not have the same colour concepts as we do". Sure enough, there are people who are colour-blind, hence it seems fair to assume that there could be a whole tribe of colour-blind people. But does the fact that all members of a group are colour-blind entail that they must have different colour concepts than we do?[1] In order to answer this question, we will have to attend to what "not having the same colour concepts as we do" means.

Here is what Wittgenstein says: "assuming they speak, e.g. English, and thus have all the English colour words, they would still use them differently than we do and would *learn* their use differently." There are several things worth mentioning about this short passage: (1) A proficient speaker of English uses the English colour words in a certain way. We would not say of someone that he or she understood the English word "red", for instance, if it was used as, say, a numeral by that person. The words of a language cannot be detached from their use. "They use the English colour words differently than we do" must therefore mean something along the lines of "They use the same symbols (signs, sounds, etc.) as we do, but in a different fashion". (2) Wittgenstein speaks of "having *all* the English colour words" and it is rather unclear what he intends to get across to his readers. While it is possible to ascribe the possession of a certain colour concept to someone, it sounds awfully strange to ascribe the possession

---

[1] Quite obviously some sort of connection between the idea of someone who is colour blind and the idea of someone who has different colour concepts is suggested to us in Wittgenstein's remark. However, the existence of such a connection ought not to be taken for granted and I am indebted to Gabriele M. Mras for encouraging me to examine the transition from one idea to the other in greater detail.

of a certain colour word or all the English colour words to someone. Furthermore, the expression "having such and such a colour word" may occur in statements like "In English we have the colour words 'honey yellow', 'vermilion', 'cream', 'ocean blue', etc." and this will mean roughly that a speaker of the language has certain words at his disposal; but we may just as well invent a new colour word if the need arises, which makes it difficult to get a clear grasp on the idea of "all the English colour words". What seems likely is that Wittgenstein wants us to assume, that these people make regular use of certain basic colour words like "red", "green", "blue", "yellow" and so on. (In §14, for instance, he goes on to speak of expressions like "reddish-green" or "yellowish-blue".) And we may anticipate a connection between his insistence that these people use "all the English colour words" and the fact that they are colour-blind. (3) Wittgenstein emphasizes that these people do not only use their colour words differently, but that they "*learn* their use differently". Given that we learn the use of a word by imitating others, it seems hardly worth mentioning and even less in need of emphasis that a different use will involve that the use of these words will be learned differently.

In order to clarify this last point, let us return to the beginning of the quoted passage once more. We were asked to imagine a tribe of colour-blind people. What is certain is that members of such a tribe will lack certain skill which we have. However, "colour-blind" may refer to different conditions of which I want to very briefly discuss three. The first is a form of colour-blindness I intend to call "property blindness" and I hope to be able to explain what I mean by that in a minute. The second way to understand the expression "colour-blind" is to think of these people as being unable to distinguish different hues so that they only discriminate between different intensities of brightness such that we may think of them as seeing nothing but shades of grey— the way things appear to us in a black-and-white movie. The third and last option I am going to discuss is that these people are either red-green-blind or yellow-blue-blind.

Let me start with the first possibility. Imagine that these people are unable to perceive colours as properties of things. What they do see, we shall assume, are what may be described as "coloured spots" or "coloured shadows". While colours are properties of things, shadows are not. An object casts a shadow. We may speak of "an object's shadow" much in the same way that we talk of "an object's colour", but a hadow is not therefore a property of an object in the sense in which a colour is a property of an object. Thus, supposing that these people have acquired a language we should be wary to say that they have colour concepts different from our own. Rather, it seems, they do not have colour concepts at all.[2] Why? Because it seems difficult to imagine how they would learn to describe their visual experience. From the assumptions made, it fol-

---

[2] Note that the scenario described thus far bears some resemblance to §2 of the "Philosophical Investigations". Remember that the builders do not use the words "block", "pillar", "slab", and "beam" to describe anything (as anything) and one may even be hesitant to say that §2 establishes a group of

lows that they will not be able to discriminate different objects (chairs, tables, etc.) by looking around. Therefore, they will lack a point of reference outside their own experience; there is nothing out there, so to speak, that has a colour about which they could come to agree. They may succeed to some extent in describing their visual experience in the way we are able to describe our feelings, but such a way of talking about what—for a lack of better words—we may call "colours" bears little resemblance to how we make use of colour words.

The reason why the suggestion is interesting, nonetheless, is the fact that we sometimes talk about colours, as if they constituted entities that exist independently of anything in the world. We make such statements as "White is the lightest colour" or "Purple is a reddish blue" and although one may understand such statements as saying "All things with property F ("being white", "being purple", etc.) also have property G ("being lighter than any non-white object", "being a reddish blue", etc.) there is a strong temptation not to understand them in this way but rather as stating a truth about certain kinds of entities; and Wittgenstein discusses the attempt to identify such entities at several points in the *Remarks on Colour*.[3] For the question at hand, however, it is unlikely that he had something like property-blindness in mind when speaking of colour-blindness. This will become clearer as we move along.

The second possibility I mentioned was to interpret "colour-blind" as meaning that these people are not able to discriminate any hues at all. Given this assumption it seems rather odd to suppose that they would use words like "red", "green", "blue", "yellow", and so on. This is an important observation and can (in part) also be made with regard to the third possibility as to what "colour-blind" may mean. If these people are either red-green-blind or yellow-blue-blind then one may wonder about their use of the words "red" and "green" or "yellow" and "blue" respectively. We may hypothetically assume at this point, that this is the direction our investigation ought to take, and see where this leads us.

---

people who can intelligibly be said to possess concepts at all. I do not want to argue for or against this view, but there are some striking similarities to the case at hand.

3 In §57 of part III Wittgenstein writes:

> "[I]f I describe a plane surface, a wallpaper, for example, by saying that it consists of pure yellow, red, blue, white and black squares, the yellow ones cannot be lighter than the white ones, and the red cannot be lighter than the yellow. This is why colours were hadows for Goethe." (RC III, 57)

The remark appears in the context of discussing the idea of a more fundamental or simple concept of colour which Wittgenstein goes on to explain by saying that "[i]t seems that one could present [this concept] either by means of small coloured elements in the field of vision, or by means of luminous points rather like stars." (RC III, 58). It is clear that in this sense colours are not taken to be properties of any object. Similar considerations can be found in §§59-61 of part I. The possibility to relate the expression "colour-blind" to the idea of coloured shadows was pointed out to me by Gabriele M. Mras.

Given that the people we are supposed to imagine are red-green-blind, it is clear that they will find it difficult to use the English words "red" and "green" the way we use them. All the more so if we are thinking of them as a secluded group of people where everyone is colour-blind; as is suggested to us by Wittgenstein's emphasis on the word "tribe". (That is to say, we are not talking about a minority of colour-blind people living amongst a majority of people who are normally sighted. Under such circumstances the colour-blind may adapt rather well to the standard use of "red" and "green", although they will from time to time make mistakes in their application of those words.) If we are thinking of a community of people where everyone is red-green-blind, it seems curious to suppose that they use two different symbols "red" and "green" at all. What use would they make of those two words? Sure enough, they may use "red" and "green" synonymously. Yet Wittgenstein writes that "if they have a foreign language, it would be difficult for us to translate their colour words into ours", which presumably would not be the case if they use the non-English words "red" and "green" interchangeably. We may then translate those words into English as "red or green". Now, whether they use those words synonymously and in such a way that "red or green" is an adequate translation will be a matter of empirical investigation. We may for example find out, that certain shades of brown, which would be called neither "red" nor "green" by us, are called "red" or "green" by them, such that "red or green" would not be an apt translation after all. Finding a suitable translation for certain words of a foreign language is by no means an easy task and it may be tempting to think that this is what Wittgenstein is driving at, when he says that "it would be difficult for us to translate their colour words into ours". Elsewhere, however, Wittgenstein states that "if they really have a different concept than I do, this must be shown by the fact that I can't quite figure out their use of words." (RC III, 123) This suggests that the difficulty, we are to concern ourselves with, is of a more fundamental kind; a kind of difficulty that arises not from a lack of information (on how to use certain words) but from a failure to recognize the rule (that governs the use of those words).

Hence, suppose for a moment that "red" and "green" are not used synonymously by these people. Given our assumption that they are red-green-blind, those words cannot possibly fulfil the same function as they do in English. They could, however, be used for another purpose such as distinguishing ripe and unripe fruit. We may then translate their use of "⌣ is red" as either "⌣ has the colour of ripe fruit" or "⌣ is ripe" and only in the former case may we consider "red" a colour word. Interestingly enough, Wittgenstein writes, that "it would be difficult for us to translate their *colour* words into ours", seemingly presupposing that their words are colour words and thereby begging the question.

Remember that the main task set before us is to find out whether the supposition of a tribe of colour blind people leads us to a coherent picture of people with colour concepts different from our own. In order to do so we will have to be justified in believing that certain words they use are colour words, while at the same time it should be true of those words, that we will find it difficult to translate them into our colour words. This puts us in a difficult position, to say the least.

Let us take a step back and suppose for a moment that we encounter a group of people who speak a language foreign to us. To be justified in saying that they are colour-blind requires that we are able to identify a large number of their words with our colour words and furthermore come to see that they are unable to mimic certain specific ways of using those colour words. Put differently, the ascription of colour-blindness requires that we are able to identify and translate a large number of colour words from their language into ours. If, however, we assume to encounter insurmountable difficulties in trying to come up with a translation for their words, it is hard to see what justification we may have for thinking that these words are colour words and even harder to see on what grounds we may believe these people to be colour-blind.

Contrariwise, given that we already know these people to be colour-blind, we will apparently be forced to conclude, that we already possess a translation manual for their colour words and forced, furthermore, to conclude, that we have discovered them to be unable to make certain distinction we are able make. All of this implies that there could not possibly have been any principal difficulties in translating their words into ours. But then it seems that the assumption of a tribe of colour-blind people could not possibly provide us with a picture of people with colour concepts different from our own.

1 The words "reddish-green" and "yellowish-blue"

Wittgenstein seems to be well aware of this point, as can be seen from the way his thought progresses from §13 to §14. Moving away from the idea of a tribe of colour-blind people, he now speaks of "people for whom it was natural to use the expressions 'reddish-green' or 'yellowish-blue' in a consistent manner". Unmistakably, the expressions "reddish-green" and "yellowish-blue" relate to red-green-blindness and yellow-blue-blindness. Yet, it is not clear, whether we ought to think of the people in §14 as colour-blind or not.

We may assume for now that they are colour-blind. Under such circumstances, given that a red-green-blind person will find it difficult to make a difference in using the words "red" and "green", we may imagine these people to use "red", "green", and "reddish-green" interchangeably. This is one possible way to use "reddish-green" in a consistent manner. However, we will not thereby have succeeded in imagining someone with colour concepts different from our own for the simple reason that it would pose no difficulty to translate those non-English words as "red or green" (for instance). Another possibility is that the expression "reddish-green" is used in such a way, that its meaning cannot be derived from the use of the words "red" and "green". Yet, if all we know of a word is that it is used in a consistent manner, we have not been told anything about the function the word fulfils in the language and so the question arises, whether "reddish-green" is a colour word.[4] (The same holds true mutatis mutandis for the words "yellow", "blue", and "yellowish-blue")

Before moving on, it should be noted that Wittgenstein speaks of people for whom it was *natural* to use certain expressions in a consistent manner. It is not immediately clear, what contrast is to be brought out by the word "natural", though. It could be that we are to compare these people with beings who found it natural to use those very expressions in an inconsistent manner. However, if a symbol is used inconsistently we ought not to call the symbol a word, much less a colour word. If, on the other hand, we compare these people to beings who did not find it natural to use those words in a consistent manner, but then discover that such a use could be established nonetheless, we may wonder what difference it makes, whether such and such a use came naturally to them or not. The important point, therefore, seems to be this: *we* do not, as a matter of fact, use these words to describe the colour of things; and it is not just the case, that such a use does not come naturally to us, but what is natural to us is not to use those expressions.

Let us return to the question of what a consistent use of "reddish-green" or "yellowish-blue" might look like. Thus far, little has been done to explain how the people we are thinking of might use those two expressions; and in §14 we are told nothing but that they "perhaps also exhibit abilities which we lack". At first glance, this may appear to suggest that we are to think of ourselves in a position similar to a colour-blind person compared to a normally sighted one. That is to say, we may come to believe that someone, who uses such expressions as "reddish-green" or "yellowish-blue" in a consistent manner, sees colours which we do not see, and in what follows I will examine this suggestion more closely.

Remember that a normally sighted person does not speak of a "reddish-green" or a "yellowish-blue" colour, while a red-green-blind person might be imagined to use "reddish-green" just because he or she will not be able to distinguish (successfully) between "red" and "green". Now, if we attempt to imagine people who see colours which we do not see, over and above the colours which we do see, we should expect them to have (at least in principle) colour words which are unknown to us. Furthermore, it seems reasonable to suppose that the combinations "reddish-green" and "yellowish-blue", which do not make sense to us, should not make sense to them either. (The fact that someone uses the expression "reddish-green" suggests, prima facie, that this person is either colour-blind or doesn't know the meaning of the words "red" and

---

4 To deduce from the fact that the expression "reddish-green" is composed of the words "red" and "green" that "reddish-green" must be a colour word, is to suppose that there is a way in which an understanding of "reddish-green" can be derived from an understanding of the words "red" and "green". If we suppress this assumption, nothing that has been said so far can be brought forward to support the idea that "reddish-green" is a colour word.

"green".) Thus, there is a tension between the idea of someone who makes consistent use of the expression "reddish-green" and the assumption that this person exhibits abilities which we lack.

There are different ways in which one may attempt to resolve this tension. We may, on the one hand, insist that these people are indeed colour-blind and that their use of "reddish-green" or "yellowish-blue" has nothing to do with the way they perceive colours. In using these expressions they may well exhibit certain abilities which we lack, but if these abilities do not relate to their perception of colours, we should be hesitant to think of the expressions "reddish-green" and "yellowish-blue" as colour words. We may, on the other hand, suppose that these people are not in fact colour-blind and explain their use of "reddish-green" by assuming their experience of colours to be (radically) different from our own, allowing for the possibility that an expression like "reddish-green" poses no difficulty for them. This suggestion has a lot going for it, given that we commonly assume our description of colours as reddish-yellow, yellowish-green, greenish-blue, etc. to be grounded in the way we experience colours and the way we experience colours to be determined by our physiology which may, as a matter of fact, have turned out to be different from what it is.

"But", says Wittgenstein, "we would still not be forced to recognize that they see colours which we do not see". A couple of things have to be noted about this. In the German original, Wittgenstein emphasises the word "colours".[5] This is especially remarkable as we would expect the emphasis to be on "colours which *we* do not see" or on "colours which we do *not* see" but not on "*colours* which we do not see"; and what is even more striking is that up to this point Wittgenstein talks about nothing but words and the use we make of those words. Rather suddenly, he now speaks of colours and our perception of colours.

Surely enough, this shift in focus is tempting. We are imagining someone who, in his or her use of the expressions "reddish-green" or "yellowish-blue", exhibits abilities which we lack. What could be more natural than thinking this person is able to see colours which we do not see? It may well be that Wittgenstein wishes to point out this temptation to us by his emphasis on "colours" and to invite us to reconsider the situation. Nonetheless, it seems like a perfectly reasonably transition to make and the claim that "we would still not be forced to recognize that they see colours which we do not see" is difficult to comprehend. However, it is advisable that we take seriously whatever Wittgenstein brings forward in order to support his claim, before jumping to conclusions. He writes: "There is, after all, no *commonly* accepted criterion for what is a colour, unless it is one of our colours."[6] The fact, that Wittgenstein here speaks of the criterion for what is a colour, may be taken as a request to consider under what

---

5 The emphasis was removed by the translators without comment.
6 The emphasis on "commonly" was added by the translators and has no counterpart in the German original.

circumstances we would say of someone, who uses the expressions "reddish-green" or "yellowish-blue" in a consistent manner, that he or she sees colours which we do not see.[7] (Incidentally, this will shift the focus of our discussion back to the use we make of certain words.)

But what is it to be one of our colours? I think the most sensible thing to say here is that a colour is one of our colours, if it can be described by a suitable combination of the six colour words "red", "green", "yellow", "blue", "black", and "white", excluding all combinations in which "red" and "green", "yellow" and "blue", or "black" and "white" appear simultaneously. That is to say, even in the unlikely event that we were to encounter people who found it natural to describe the colour of certain things as "reddish-green" or "yellowish-blue" and it is furthermore discovered that these people have a different physiology, we would still not be able to give sense to the expressions "reddish-green" or "yellowish-blue". But why should it be necessary that we be able to do so? If that were a condition on our being justified in believing that someone sees colours we do not see, it is difficult to imagine, how we should ever come to be in a position to say such a thing of anyone. That much is true. However, we are not concerned with the general question, under what circumstances the belief, that someone sees colours which we do not see, could be justified. Rather, we are considering, whether there are good reasons for believing that someone sees colours which we do not see, if this person makes consistent use of the expression "reddish-green" or "yellowish-blue".

Remember that we were invited in §14 to imagine a group of people, who exhibit certain abilities which we lack. Certainly, we could simply declare that they are able to see colours we do not see. However, such a declaration has little significance for the question at hand. What we would like to know is whether we could, on the basis of observing their behaviour, come to be justified in believing that these people see colours which we do not see. Given this it seems plausible to suggest that we would need to come up with certain tasks, similar to the Ishihara test for colour blindness; and by describing those tasks and the way these people perform in them, we will have given reasons for believing that they see colours which we do not see. How good a justification is this, though?

---

[7] In making this suggestion I rely rather heavily on Stanley Cavell's discussion of Wittgenstein's use of the word "criterion", which can be found in the first chapter of his wonderful book "The Claim of Reason" to which I would like to refer the reader at this point.

2 The description of consistent behaviour

Suppose we ask someone of whom we are told that he or she uses "reddish-green" or "yellowish-blue" in a consistent manner, what those expressions mean. We may expect to be shown several coloured objects. Quite clearly, though, this will not bring us closer to an understanding of those words. Just as it will be no help to a red-green-blind person, if we point out several red and green objects in order to explain the difference between "red" and "green".[8] Furthermore, coming to believe that a person uses a certain word in a consistent manner—either by being told so or by controlled observation—does not ensure that the word is indeed used consistently. If an expression is used consistently, there must be rules for its use; so much is entailed in the definition of "consistent" as "of a regularly occurring, dependable nature", "free from contradiction" and "free from variation". However, a red-green-blind person will be in no position to judge whether a normally sighted person uses the word "red" in a consistent manner, and the same is true for us with respect to someone of whom it is thought that he or she uses "reddish-green" in a consistent manner.

To clarify this point let us imagine a being that speaks a language foreign to us and that some effort has already been made to come up with a theory, that allows us to translate certain of those words into English. Assume, furthermore, that "ajel" and "benu" are two words for which our best translation suggest that "ajel" means red and "benu" means green. What should we say, if the being we are imagining sometimes combines the words "ajel" and "benu" when asked for the colour of things? We may then and there feel some pressure to admit that those two words do not translate as "red" and "green" as smoothly as initially thought. Without a doubt, one possible explanation may be that such a being sees colours which we do not see. But it is far from clear that this is the best explanation available. It might just as well be that the combination of "ajel" and "benu" constitutes an irregularity in the foreign language. Can this possibility be ruled out somehow? Suppose we discover that the being we are imagining is able to distinguish certain chemical substances merely by looking at them and that it describes the colour of one of the substance by a combination of "ajel" and "benu". Even in such a scenario and granting that our translation for "ajel" and "benu" is flawless for all cases in which those two words are not combined, we may be hesitant to admit that we have imagined a being, able to see colours that we do

---

8 Tobias Rosefeldt has called attention to a very closely related point in an article titled "Was Mary nicht konnte und was sie nicht wußte" from 2001. Rosefeldt convincingly argues that someone, who is able to use such expressions as "⌣ is red" in accordance with us, does not necessarily understand "red" the way we do. To understand a colour word consists essentially in being able to decide of a thing what colour it has without already being acquainted with it. Someone who is merely able to tell the colour of a thing on the basis of a memorised list does not know the colour of things, because he or she lacks a proper understanding of those terms.

not see. This has to do with the fact, that the combination of "red" and "green" does not make sense in the way that "reddish-yellow", "yellowish-green", "greenish-blue", etc. make sense. The description of a colour as reddish-green or yellowish-blue is ultimately incomprehensible to us. But, we are tempted to ask, could such a description not be comprehensible to us if certain general facts of nature were different? Well, if there were a scenario under which "reddish-green" made sense, we would not be talking about the English words "red" and "green" anymore, raising anew the question whether those foreign words are colour words.

What is shown by these considerations is that there is something worrying about the claim that the expressions "reddish-green" and "yellowish-blue" do not denote any colours for us. It would be more appropriate to say that these terms do not denote anything at all. The temptation to speak of colours different from our own lies, I think, in the conviction that it is clear enough what "colours, different from the ones we know" means. In most situations we are able to decide whether two colours are different from each other. This can easily lead to the assumption that the expression "a different colour" has a clear sense even when we are not speaking of red, green, blue, etc.; as if it were enough to say "You know what it means to say that red is a different colour from green. In this sense I am using the expression 'different colour' when I speak of colours, different from all the colours we know." But while there is a certain practice of comparing colours with each other, there is no practice of comparing all of our colours with something that is not one of our colours. And our usual way of comparing colours provides no guidance as to how such an extension of our use of colour words may be carried out. Wittgenstein suspects that one source for such misunderstandings lies in the conception of colours as things rather than properties.

> "'The colours' are not things that have definite properties, so that one could straight off look for or imagine colours that we don't yet know, *or* imagine someone who knows different ones than we do. It is quite possible that, under certain circumstances, we would say that people know colours that we don't know, but we are not forced to say this, for there is no indication as to what we should regard as adequate analogies to our colours, in order to be able to say it. This is like the case in which we speak of infra-red 'light'; there is a good reason for doing it, but we can also call it a misuse." (RC III, 127a)

According to this, there may be situations in which there are good reasons for speaking of "colours, different from the ones we know", but we may also consider it a misuse of the word "colours"; especially so, if these alleged colours are described to us as reddish-green or yellowish-blue.

References

Cavell, S.: *The Claim of Reason*. Oxford: Oxford University Press, 1979.
Rosefeldt, T.: Was Mary nicht konnte...und was sie nicht wußte. Wittgensteins Beitrag zur Qualia Debatte. In: Puhl, K./ Haller, R. (Eds.), *Wittgenstein and the future of philosophy – a reassessment after 50 years, 24 International Wittgenstein Symposium*, p. 247–253, Wien: ÖBV + HPT, 2001.

Herbert Hrachovec
Reddish Green

Imagine an autumn leaf changing its color between green and red. The color of some of your garden's roses is advertised as bluish white, so you describe the color of the autumn leaf as reddish green. (The pattern is familiar from referring to a music style as "bluegrass", i.e. a mix of blues and country music.) You are surprised when someone tells you that there is no reddish green on account of the fact that those are (according to common doctrine) complementary colors. They cannot blend into each other, just like you cannot simultaneously take the directions right and left at an intersection. Calling something reddish green violates received opinion, disregarding the color scheme governing much of ordinary discourse and hence losing control as far as color terms are concerned.

The first approach is *ad hoc* pragmatic, the second one seems to impose a pattern upon the episodes just mentioned. This paper will trace Ludwig Wittgenstein's development from a distinctly normative view to the pragmatic leanings of his later work. It will, however, refine the usual account of Wittgenstein's eventual rejection of his *Tractatus* views. It can be shown that empiricism is built into the framework of the *Tractatus*' ontology. And the conceptual concerns of this early book can, on the other hand, still be detected in Wittgenstein's seemingly open ended later writings. These claims cannot be defended as a general thesis here. They will be introduced by following the philosopher's struggle with the issue of reddish green, which turns up at various places in his *Nachlass*. It is well known that closer attention to the grammar of color terms has played a crucial role in Wittgenstein's rejection of essential *Tractatus* doctrines. What are the relevant propositions and why have they been put into question?

1 Logical form versus empirical research

In an entry in his diary from 22nd January 1915 Wittgenstein is setting forward an ambitious agenda. He describes his overall task as explaining the essence of *the* sentence, i.e. of all facts that can be pictured. "Das Wesen allen Seins angeben."[1] Or, viewed from a slightly different angle:

---

[1] "In giving the nature of all being."

> "Das große Problem, um welches sich alles dreht, was ich schreibe, ist: Ist, a priori, eine Ordnung in der Welt, und wenn ja, worin besteht sie? (NB, 1st June 1915)"[2]

The attempt to delineate an ordered view of the world achievable prior to actually taking empirical investigations into account is faced with considerable difficulties, many of which can be traced back to Wittgenstein's notebooks. A crucial component of a world, orderly conceived, are "Gegenstände" (things) and surely their availability as building blocks has to be presupposed in advance of their *de facto* occurence (cf. NB, 16th June 1915). But a characteristic dilemma arises here. Things, according to ordinary parlance, can be quite complex. A well-founded world, however, has to be built up from *basic* givens, simple things that cannot be further analyzed. It does not seem feasible to a *apriori* predict what these simples are going to be.

Wittgenstein spends considerable time to become clear about this point. The upshot is that things enter into states of affairs by virtue of their logical form. "'Die Uhr *sitzt* auf dem Tisch' ist sinnlos."[3] Names are conditioned by what Wittgenstein calls the "logical kind of the object" (NB, 22nd June 1915) and this includes complex as well as simple things:

> "Und das ist klar, daß der Gegenstand eine bestimmte logische Art haben muß, er ist so zusammengesetzt oder so einfach als er eben ist."[4] (NB, 22nd June 1915)

In view of the quest for the essence of being this account is generalized in the notebooks on 11th July 1916 and finds its way into the *Tractatus*:

> "Wenn die Gegenstände gegeben sind, so sind uns damit auch schon *alle* Gegenstände gegeben. Wenn die Elementarsätze gegeben sind, so sind damit auch *alle* Elementarsätze gegeben."[5] (TLP, 5.524)

The big if is, obviously, how we could come to know the internal shape of things concatenated in elementary propositions as a matter of logical insight – a task mandated by Wittgenstein's metaphysical ambitions. He is juxtaposing two conflicting intuitions:
1. Matters of logic have to be settled independently of experience. (TLP, 5.551ff.)
2. Empirical reality manifests itself withing the totality of elementary propositions. (TLP, 5.5561)

---

2 "The great problem round which everything that I write turns is: Is there an order in the world apriori, and if so what does it consist in?"
3 "The watch is sitting on the table' is senseless!" (NB, 22nd June 1915)
4 "And it is clear that the object must be of a particular logical kind, it just is as complex, or as simple, as it is."
5 "If the objects are given, therewith are *all* objects also given. If the elementary propositions are given, then therewith *all* elementary propositions are also given."

It is by no means clear how a "logical form" of things determining their fit in states of affairs could be dependent on empirical circumstances. Wittgenstein summarizes the problem in a remarkable aphorism:

> "Es soll sich a priori angeben lassen, ob ich z. B. in die Lage kommen kann, etwas mit dem Zeichen einer 27-stelligen Relation bezeichnen zu müssen."[6] (TLP, 5.5541)

In order to capture the underlying structure of the world we need a logical syntax pre-arranged to fulfil the task of expressing facts that can, precisely, *not* be foreseen by logic. Wittgenstein's attempted resolution of this issue cannot be discussed here. The point is rather that already in his *Tractatus* period he faces the problem of the *provenance* of (knowledge about) logical forms designed to govern empirical descriptions. The question whether a 27-ary term should be a meaningful expression has to be settled *apriori*. If we apply this *dictum* to colour terms the difficulty becomes obvious. The logic of pronouncing a patch as possessing one colour does not simply consist in this patch either exhibiting or not exhibiting said colour. Given the colour spectrum the affirmation of a colour sentence immediately implies a multitude of negative assertions not included in a two-valued approach. This observation led Wittgenstein to drop the requirement of mutual independence of elementary propositions.[7] Returning to our initial scenario of describing an autumn leave we can recognize the impasse between pragmatic and normative forms of discourse as being an instance of a more general tension. What are the rules governing the use of "reddish green"? Can we combine expressions referring to these colours to indicate a mixtum compositum, or are we obliged to reject this option by arguments referring to the conceptual incompatibility of those respective predicates?

---

[6] "It should be possible to decide a priori whether, for example, I can get into a situation in which I need to symbolize with a sign of a 27-termed relation."
[7] Cf. MS 112, p. 134r:

> "Man kann den Satz 'dieser Ort ist jetzt rot' (oder 'dieser Kreis ist jetzt rot' etc.) einen Elementarsatz nennen, wenn man damit sagen will daß er weder eine Wahrheitsfunktion anderer Sätze ist noch als solche definiert ist. [...] Aus 'a ist jetzt rot' folgt aber 'a ist jetzt nicht grün' und die Elementarsätze in diesem Sinn sind also nicht voneinander unabhängig, wie die Elementarsätze in meinem seinerzeit beschriebenen Kalkül von dem ich annahm, der ganze Gebrauch der Sätze müsse sich auf ihn zurückführen lassen; – verleitet durch einen falschen Begriff von diesem 'Zurückführen'."

2 Synthetic knowledge a priori

The complementary character of red and green plays an important role in challenging the *Tractatus*' atomism.

> "From 'a is not red now' it can be followed that 'a is not green now' and elementary propositions are therefore not independent from each other like elementary propositions within my erstwhile constructed calculus [...]."[8] (MS 112, p. 134r)

Wittgenstein takes up the point much later, in 1946, still in the normative voice implied by his reference to a calculus.

> "'Ein rötliches Grün gibt es nicht' ist den Sätzen verwandt, die wir als Axiome in der Mathematik gebrauchen."[9] (MS 133, p. 25r)

One similarity consists in both mathematical axioms and the given proposition not being derived from observation. They are stipulations and their validity rests on construction rather than empirical evidence. Yet, Wittgenstein's comparison is somewhat surprising in view of the fact that colours are, supposedly, sense impressions external to our cognitive machinery. The *Tractatus* does not provide the reader with any means to understand how an axiomatic stipulation may match with empirical validation. This is where Wittgenstein tentatively enters into the realm of grammatical explorations which are reminiscent of the style of transcendental arguments.

It is not just, in fact, the case that red and green do not mix at any given location. The force of this statement might be considered equal to an assertion like "Mahatma Gandhi never met Dwight Eisenhower in their entire lifetime." The mix is *impossible* in a certain system. Wittgenstein's claim points back to the *Tractatus* pronouncement that a definitive language will by itself rule out nonsensical expressions. This (kind of) postulate cannot, however, easily survive the impact of Wittgenstein's rejection of elementary propositions. Determining the logic of the colour spectrum is, remember Wittgenstein's remark, like knowing in advance whether there will be any need for an expression referring to a 27-ary relation. Wittgenstein seems to be aware of the difficulty and continues his entry with the following, somewhat precarious, reflections:

> "Wir sagen nun: 'wir gebrauchen die Wörter 'rot' und 'grün' in solcher Weise, daß es als sinnlos gilt (kontradiktorisch ist) zu sagen, am selben Ort sei zu gleicher Zeit rot und grün'. Und dies ist natürlich ein Satz."(MS 113, p. 29v)[10]

---

8 German original in previous footnote. All translations from Wittgenstein's manuscripts are by H. H.
9 "'There is no reddish green' is related to those sentences we use as axioms in mathematics."
10 "And now we say: we use the words red and green in such a way that it counts as meaningless (as contradictory) to assert that red and green are at the same place at the same time."

This quote mixes several conflicting approaches. Contradiction is easily defined in formal logic. It is, however, by no means obvious that it can be equated with lack of meaning, as Wittgenstein seems to suggest. No reference to meaning is necessary to recognize a contradiction holding between a sentence and its negation. The kind of "contradiction" Wittgenstein is invoking seems to rest on semantic incompatibility. Model theory can formalize the permitted scope of predicates and consequently reconstruct e.g. exclusions within a colour scheme. But this is not the kind of regiment Wittgenstein is looking for.

Wittgenstein is, in fact, not just positing the semantic incompatibility between red and green, he additionally introduces the pragmatics of this setup. Our *use* of those colour expressions blocks us from asserting certain sentences. The meaning of terms is derived from the overall behaviour of the language community. The said contradiction is, according to this account, the self-imposed rule of a language game. The inconspicuous quote under consideration contains, as it turns out, the whole range of Wittgensteinian options in dealing with the issue of philosophical grammar. One way to read it is to emphasize its "deconstructive" character. A forbidding array of prohibitions has been assembled in the *Tractatus* – they can now be seen as metaphysical superstructure erected on the foundation of certain social practices. The issue of red-green incompatibility seems to collapse into a discrepancy between various linguistic attitudes. But there is an alternative reading, *preserving* the tension between *apriori* motives and empirical outlook that we started with. This reading registers the nuances between a logical calculus, accounts of meaning and theories of use, while at the same time remaining aware of their systematic peculiarities. It can be shown that Wittgenstein was struggling with those peculiarities up well into his late writings on colour.[11]

A non-technical way to summarize the bundle of intuitions surrounding "reddish green" is:

> "Man kann aber auch so sagen: Wenn ich das Produkt zweier Sätze bilden kann, so können sie nicht die Sinne haben 'a ist rot' und 'a ist grün'. [...] Die beiden Sätze kollidieren im Gegenstand."[12] (MS 106, p. 87f.)

Taken at face value the last sentence *does not* defuse the collision by introducing negotiable rules but rather pins down the logico-ontological challenge: the incompatibility of sentences is grounded in real world things. This insight can, using a more traditional philosophical vocabulary, be described as synthetic knowledge *apriori*. The compatibility in question is outruled in advance. We know that it cannot *occur* since

---

11 The issue of *apriori* knowledge with regard to colour incompatibilities has, among others, been discussed by Jacques Bouveresse in his "Wittgenstein's Answer to 'What is Colour?'".

12 "One can also put it this way: If I can form the product of two sentences they cannot have the meanings 'a is red' and' and 'a is green'. [...] Those two sentences collide with respect to their object." (Literally: "collide in the object")

we cannot even (legitimately) *conceptualize* it. It has, by now, become clear that one-sentence pronouncements merely serve to highlight the underlying problem, as another quote from a 1932 manuscript demonstrates.

Wittgenstein draws a distinction between (1) "No book is lying here" and (2) "The colours red and green can simultaneously occupy one spatial position". A book could be found at this (currently empty) spot; red, on the other hand, *cannot* occur at green patches. The latter option is, as discussed earlier, excluded by logical semantics, i.e. philosophical grammar. But difficulties are just starting here.

> "Aber, wenn der Satz dadurch sinnvoll wird, daß er mit den grammatischen Regeln in Einklang ist, so machen wir eben die Regel, die den Satz 'rot und grün sind zugleich an diesem Fleck' zuläßt. Gut; aber damit ist nun die Grammatik dieses Ausdrucks noch nicht festgelegt. Es müssen erst noch weitere Bestimmungen darüber getroffen werden, wie ein solcher Satz zu gebrauchen ist; wie er z.B. verifiziert wird." (MS 114, p. 118)[13]

If the red-green incompatibility were closely similar to mathematical axioms we would be free to lay down alternative rules and to allow co-occurrence of those colours. This would put Wittgenstein on the side of pragmatism, providing methodological cover for multiple discursive contexts that exhibit probably mutually exclusive points of departure. Wittgenstein *does not* hold this position in the above quote. He demands an additional move to fix the expression's grammar and introduces its *use*, in particular its method of verification, as determining factor. Now, this move can only support the categorical incompatibility when this very incompatibility is written into the language we use to articulate verifiable statements. There is no point in trying to verify a statement according to which a co-occurrence of red and green is the case (or not the case), as long as it remains undecided whether we are even allowed to entertain this (Fregean) thought.

One way to go from here is to drop the strong suggestion of a pre-determination of meaning and this is, indeed, what Wittgenstein's later writings are often taken to do. But this is by no means what his remarks show. On closer inspection Wittgenstein can be seen to tentatively switch sides on a number of occasions. He never completely abandons his initial in principle rejection of a red-green mix, yet, on the other hand, is characteristically inventive in coming up with scenarios contradicting this orthodoxy. One of his favourite examples in discussing the issue actually is autumn leaves. He mentions them as early as 1936 (MS 115) and as late as 1949 (MS 171). One may imagine a group of men perceiving their surroundings as a continuous transition between

---

[13] "But, if a sentence becomes meaningful by corresponding to the grammatical rules we can just lay down the rule which allows the sentence claiming that red and green are simultaneously at this patch. Fine; but the grammar of this expression is not yet fixed by this move. Additional provisions about the use of such a sentence have to be made; e.g. how it is to be verified."

red and green "und zwar so wie wir es im Herbst an manchen Blättern sehen"[14] (MS 115, p. 238):

> "Ich stelle mir vor, es gibt nur einen Ton von Rot und von Grün. Die beiden gehen in der Natur (wie im Herbst in gewissen Blättern) immer ineinander über."[15] (MS 137, p. 100b)

> "Nichts ist so gewöhnlich wie die Farbe rötlichgrün; denn nichts ist gewöhnlicher als der Übergang vom Grün des Blattes in's Rote."[16] (MS 171, p. 14)

Wittgenstein's remarks make a strong case for discarding attempts to develop a constitutive account of colour space in general. They are, however, counterbalanced by attempts to avoid relativistic conclusions that might be drawn from thought experiments imagining various possible forms of life.

3 Limits to change

The contentious issue of reddish green is not resolved in Wittgenstein's oeuvre. Rather than offering a solution Wittgenstein involves himself in pointed dialogues, switching sides more than once. An instructive series of remarks appears in manuscript 137, selected parts of which he chose to include in typescript 232. This dictation was published as second volume of "Bemerkungen über die Philosophie der Psychologie" by G. H. von Wright and H. Nyman in 1980. It contains a multi-faceted discussion of the issue at hand. Characteristically, Wittgenstein is cross-examining himself. One voice, let us call it the normativist, points out that concepts are accompanied with certain sentences that are incontestably part of their meaning. "There cannot be a regular digon" (MS 137, p. 5a). Someone who claims otherwise does literally not know what he is talking about. This example from mathematics is taken as a guide for considerations of the colour spectrum. Significant similarities *and* dissimilarities can be observed. There is, on the one hand, a strong intuition that someone who attempts to mix complementary colours is missing a crucial point of our common use of colour terms. Yet, colour sensations are not on a par with formal constructions. We cannot infer exclusion of alternatives from sense data. Colours are qualitative givens over and above the conceptual structures we impose upon them.

At this turn the opposing party, the pragmatist, is making her point. An assertion about numbers or geometry might be obvious to a mathematician, but perceptions

---
[14] "the way we see some leaves in autumn"
[15] "I imagine that there is just one tint of red and of green. In nature they always shade into each other (like certain leaves in autumn)."
[16] "There is nothing more common than the color reddish green; because nothing is more common than the transition from a leave's green into red."

are different. Normative structure is wedded to a conceptual framework which is not inherent in sense impressions and of which more than one version is permissible.

> "Aber haben denn Farben eine Struktur? Die Anwendung des Farbworts hat eine. Und insofern hat der Begriff eine."[17] (MS 137, p. 5b)

The pragmatist's position does, however, not satisfy Wittgenstein. Its constructivism seems too easy. "Zu sagen 'Es liegt an den Begriffen klingt wie: Nichts leichter, als andere Begriffe zu schmieden'."[18] (MS 137, p. 5b) The pendulum swings back. "Zwischen grün und rot, will ich sagen, sei eine geometrische Leere, nicht eine physikalische."[19] (MS 137, p. 5b) Can one have it both ways? Is it possible to reconcile two apparently conflicting intuitions, namely the extra-empirical cogency of categories constitutive of our world view and, on the other hand, the ultimate role of empirical evidence as an arbiter of epistemological claims? Wittgenstein's move, when pushed into such corners, often is to step sideways, i.e. to introduce a comparable case in order to compare and differentiate. At this particular juncture it is human nourishment.

Our handling of the red-green dichotomy is neither a matter of logic nor of arbitrariness.

> "Ist denn, daß wir die Dinge in dieser Weise (miteinander) vergleichen, sie so im Gebrauch zusammennehmen, ist dies denn willkürlich? Nicht mehr, als daß wir uns von diesem und nicht von jenem nähren."[20] (MS 137, p. 5b)

The opposing "voices" can easily be recognized in this scenario. A pragmatist might claim victory since, obviously, humans eat and reject to eat, quite different stuff. But this does not hold for, e.g. poisonous mushrooms the normativist is likely to reply. Some seemingly edible substances destroy the recipient biological system. While this certainly is an empirical fact, more importantly it serves as a kind of definition of this very system. We are back at "begriffsbestimmende Sätze" and their precarious relation to experience. Wittgenstein's considerations are, in fact, a sequence of variations on this overarching theme.

Is our system of colour terms arbitrary? One of Wittgenstein's responses is awkwardly evasive. "Ja und nein. Es ist mit Willkürlichem verwandt und mit Nicht-Willkürlichem."[21] (MS 137, p. 6a; also TS 232, p. 426) The crucial question, unanswered by this conjunction, concerns the specifics of this arbitrariness. The natural history

---

[17] "But do colors, then, possess structure? The use of the color word has one. And the concept, in this respect, has one."
[18] "To say it is a matter of concepts sounds like: nothing is easier than forging different concepts."
[19] "Between green and red, I want to say, there is a geometrical void, not a physical one."
[20] "Is the fact, then, that we compare things (to each other) in this way, summon them up in use, is this, then, arbitrary? Not more so than we nourish ourselves from this and not from that."
[21] "Yes and no. It is related to the arbitrary and to the non-arbitrary."

of colours is invoked as supporting the pre-established harmony of the colour octahedron with his real life employment (MS 137, p. 7b). But Wittgenstein recognizes that the binding quality of the scheme cannot be claimed to represent a natural phenomenon. Painting in ancient Egypt is not an arbitrary style, but neither is it given *a priori* (MS 137, p. 8a). One more variation: a room's furniture is designed to fit human needs, pragmatism is right about this. Yet, its purpose could not be fulfilled unless this equipment exhibited certain characteristics of robustness and convenience (MS 137, p.8b). A realist thread runs through the dialectic we are discussing:

> "Ja aber hat denn die Natur hier gar nichts mitzureden?! Doch – Nur macht sie sich auf andere Weise hörbar. 'Irgendwo wirst Du doch an Existenz und Nichtexistenz anrennen!' – Das heißt aber doch an Tatsachen, nicht an Begriffe."[22] (MS 137, p. 7b)

An editorial detail sheds some light upon the character of these exertions. Wittgenstein inserts an aside into the sequence of remarks dealing with reddish green. Just one sentence, partly exasperation and partly self-consolation.

> "Wenn die Türe nach innen aufgeht, und ich nicht daran denke, sie könnte so aufgehn, so bin ich eingesperrt."[23] (MS 137, p.8a)

This echoes the better known variation of the same idea at MS 125, p. 57v and raises the question whether there is a Wittgensteinian way out of the impasse.

Wittgenstein's discussion remains inconclusive. The following attempt to push the matter one step further turns to one more *Nachlass* entry which appears in a notebook from 1950 and might help finding an escape exit. Wittgenstein has, for the most part of MS 137, p. 5a-9a weighed the difference between *a priori* and empirical intuitions. Another approach is only mentioned in passing, namely a no-nonsense "ethnocentric" stance.

> "Die Leute kennen ein Rötlichgrün. Aber es *gibt* doch gar keins! – Welcher sonderbare Satz. – (Wie weißt Du's nur?)"[24] (MS 137, p. 7a)

Rather than try and trace the legitimacy of talk about reddish green to axioms and/or evidence this quote draws the line between established language use and dissidents. Reddish green is simply ruled out. Our tool-set does not provide a handle to deal with this so-called colour. Constitutive categories and supporting experience are conve-

---
[22] "Yes, but does Nature have no say at all here?! Yes. It is just that she makes herself heard in a different way. Somewhere you will be running up against existence and non-existence! Which means against facts, not concepts."
[23] "If the door opens to the inside and I do not think that it might open this way I am locked in."
[24] "People know of a reddish green. But there *is* none! What a peculiar sentence. (How could you know?)"

niently matched in everyday procedures. Wittgenstein is, of course, far too sophisticated to let the matter rest here. Given those circumstances the pronouncement denying the existence of reddish green is not comparable to statements asserting or denying the presence of some colour at a certain place. We cannot deny the presence of "colour" within our color spectrum while failing to consider it a colour at all. This line of argument abandons the struggle with the form-content distinction looming behind the manifest problem and rather stages a conflict between an operative synthesis of dealing-with-the-world and moments of stress, when this synthesis' internal expressive resources are overexposed in attempts to cross a boundary they are not developed to trespass.

Wittgenstein discusses this unilateral account at the limit of language mastery. If meaning is (closely linked to) use one must be able to teach an expression's use in order to convey its meaning. Familiarity with a concept has to amount to being able to introduce and apply it in suitable ways, i.e. in circumstances validated by the appropriate language games. But we are working under the assumption that, from the point of view of a certain community, there is no access to this kind of explanation.

> "Aber kann ich doch die Praxis von Leuten beschreiben, die einen Begriff haben, z.B. 'rötlichgrün', den wir nicht besitzen? – Ich kann diese Praxis doch jedenfalls niemand lehren."[25] (MS 173, p. 28r)

The world of black and white photography, for example, does not offer any help in approaching the issue of reddish green. Speakers are at a loss trying to use the resources of their language in order to introduce terms violating the patterns of this very language. One cannot have it both ways, sharing a determinate linguistic environment with a "like-minded" group *and* referring to outside phenomena excluded by the group's consensus.

> "Kann ich denn auch nur sagen: 'Diese Leute nennen dies (ein Braun etwa) rötlichgrün'? Wäre es dann eben nur ein andres Wort für etwas, wofür auch ich eins habe?"[26] (MS 173, p. 28r)

The unilateralist's dilemma is different from the give and take between the normativist and the pragmatist. She finds herself attempting to adjudicate questions she lacks, by definition, the means to adequately handle. The discussion in section 2 of this paper involved conflicting claims about the source of validity of a given cognitive setup. The present quote, in contrast, deals with the difficulty of understanding something distinctly different from an established language game.

---

[25] "But can I describe the practice of people who have a concept, for instance of 'reddish green', which we do not have? – In any case, I cannot teach this practice to anyone."
[26] "Could I even say: These people call this (e.g. some Brown) reddish green? Wouldn't this be just another word for something I myself have a word for?"

Wittgenstein is drawing a sharp line between the rules governing a linguistic community and outside interference. A touch of *Tractarian* once-and-forever brinkmanship can be detected in his phrasing ("Could I even say: ... "). Meaningful sentences have to comply to logical form. But, as we have seen, this presupposition cannot be upheld in the light of Wittgenstein's subsequent deliberations. The understanding inherent in a group's communication cannot be atemporal and *a priori*. Learning crucially depends on one's abilities to enlarge the scope of given verbal and behavioral patterns, hence the popular view that Wittgenstein turned to pragmatism once he began to reject his rigorous early demarcations. The short quote under consideration here reveals a more complicated picture. Wittgenstein starts in the assertive mood. We refer to colours by means of our vocabulary. Assume a coloured patch is brown and this very patch is called "reddish green" in a different idiom. There are, if we stick to strict unilateralism, two options. Either the foreign term translates to "brown", or we do not know what the strangers are talking about.

This is not the end of the story for Wittgenstein in 1950, though. He adds a seemingly uncontroversial, but nevertheless far-reaching remark.

> "Wenn sie wirklich einen andern Begriff haben als ich, so muß sich das darin zeigen, daß ich mich in ihrem Wortgebrauch nicht ganz auskenne."[27] (MS 173, p. 28r-28v)

The tiebreaker in the standoff between ethnocentricity and relativism is *partial* confusion concerning the interlocutor's use of terms. Certain parts of any existing colour-related discourse have to be recognizably similar to a deviant use of terms for understanding to even begin going beyond its initial borders. It seems that a shared concept of red can be universally discerned wherever colour is at stake.[28] Now, this homogeneity can break apart in more complicated cases. As a shorthand description we might describe this as "Our colour concepts fail to apply" and start to worry about incommensurability. But, given the circumstances, this response is deceptively superficial. Ordinary cases of discursive misunderstandings between cultures arise at certain junctures *recognizable as such* with respect to common background assumptions. The unilateral stance turns into an universalist outlook as far as those commonalities are concerned. And it reaches its limits at particular critical checkpoints. It is not a dialectic between form and content, but a superposition of "business as usual" and linguistic disturbance.

According to these considerations the appropriate point of departure is neither the all-pervasive power of (one?) form of life[29], nor the realm of practical expediency constantly shifting. Human interaction is based on occasional, fragile consensus. This

---

[27] "If they really have a concept different from me it has to show in that I cannot completely know my way around their use of words."
[28] Cf. Berlin & Kay 1969. For an introduction in recent research see Kay & Maffi 2011.
[29] Cf. Hrachovec 2009

turns into a unison of voices or a communication breakdown in limiting cases. A resolution of the problem of incommensurable language games is implied by this. Partial confusion concerning a concept is not always a harmless event. A small deviation between language use may, at times, make all the difference, including acceptance or rejection of an entire discursive formation. Think of theological controversies about transubstantiation resulting in religious wars, or, more down to earth, the operation of a password to access an internet site. Mistyping a single letter is all it needs to exclude a person from participation. The lesson to be learned for "reddish green" is the following one: the worry about prevailing conceptual or, alternatively, empirical determination of "the nature of colour" (MS 137, p. 7b) is an infelicitous starting point.

A better way to look at the controversy is to begin from within a language game disclosing the world in one characteristic way and prepare for eventual irritations. Disturbing situations are not to be explained by reference to unconceptualized "things in themselves", but by closer examination of the discrepancies between the entire linguistico-behavioral contexts. There will be occasions when the term "reddish green" can serve perfectly well as the description of a composite colour, yet everything might occasionally hinge on the need to emphasize the difference between complex terms like "bluish white" and "reddish green". One concluding caveat against the temptation to regard this as an essentially relativistic advice: Wittgenstein concludes his remark MS 173, p. 28v on a note of self-doubt. His later writings contain many cases of Wittgenstein imagining unexpected scenarios, stretching the meaning of formerly well-understood terms. One might get the impression that his concern with philosophical grammar gets lost amongst the micro-management of subtle linguistic nuances. But consider this quote, reigning in his imagination:

> "Ich habe aber doch immer wieder gesagt, man könne sich denken, daß unsre Begriffe anders wären, als sie sind. War das alles Unsinn?"[30] (MS 173, p. 28v)

Wittgenstein is too severe in judging himself, as usual. But there *are* limits to changing concepts. Sometimes we lose our grip on the use of presumed signifiers, leaving us tongue-tied.

References

Berlin, B. & Kay, P.: *Basic Color Terms: Their Universality and Evolution*. CSLI Publications, 1969, 1999.

Bouveresse, J.: Wittgenstein's Answer to 'What is Colour?'. In: Moyal-Sharrock, D. (Ed.), *The Third Wittgenstein. The Post-Investigation Works.*, p. 177–192, Ashgate, 2004.

---

[30] "But I said again and again that one could imagine our concepts to be different from how they are. Was all of this nonsense?"

Hrachovec, H.: Formvollendet? Nein danke! In: Denker, C. W. (Ed.), *Lebensform Wittgenstein. Bilder und Begriffe.*, p. 21–24, Passagen, 2009.

Kay, L., P. & Maffi: *Colour Terms*, Kapitel Available online: The World Atlas of Language Structures. http://wals.info/chapter/132, 2011.

Martin Kusch
Wittgenstein as a Commentator on the Psychology and Anthropology of Colour

1 Introduction

As is well known, Wittgenstein had a life-long interest in the philosophy of colour, from the *Tractatus* all the way to the last notebooks that were posthumously published as two books, *Remarks on Colour* and *On Certainty*. Moreover, Wittgenstein's various reflections of the perception and classification of colours have already been analyzed by a number of influential interpreters. These interpreters have often sought to illuminate Wittgenstein's views by relating them to other, earlier treatments of phenomena of colour, for example those written by Georg Christoph Lichtenberg (1742-1799), Johann Wolfgang von Goethe (1749-1832), Philipp Otto Runge (1777-1810), Arthur Schopenhauer (1788-1860), Franz Clemens Brentano (1838-1917), or David Katz (1884-1953).[1]

One aim of my paper is to add a new "foil" to this list: I want to make plausible that a number of Wittgenstein's remarks on colour are responses to late-nine-teenth- and early-twentieth-century British and American work on the psychology and anthropology of colour. I am not the first to put forward this idea – it is mentioned in a recent paper by the historian of science Simon Schaffer (2010: 279). But Schaffer's comment is brief, and he provides only little evidence. So there remains plenty for me to do.

I have a second aim, too. I want to argue that Wittgenstein's comments are still of systematic interest today. The link between the historical thesis and the systematic concern is established by the fact that a very influential body of contemporary work in the anthropology of colour is strongly influenced by the early British work. Presumably, if Wittgenstein's comments work as criticism of the latter, it will also weaken the appeal of the former.

My paper falls into three parts. Section 2 gives an introduction to the relevant psychological and anthropological studies. Section 3 situates some of Wittgenstein's comments vis-à-vis these investigations. Chapter 4 summarises my observations.

---

[1] Cf. Bouveresse 2004, Brenner 1982, Lee 1999, 2005, McGinn 1991, Rove 1991, Vendler 1995, and Westphal 1987.

2 The Psychology and Anthropology of Colour

In 1898-99 a number of British psychologists and anthropologists went on a Cambridge-led expedition to the islands in the Torres Strait (between Australia and Papua New Guinea) to investigate the islanders from various anthropological, psychological, physiological and linguistic perspectives. The main results were later published as the *Reports of the Cambridge anthropological expedition to Torres Straits* (1901-1935). The expedition and its results are generally regarded as a landmark event in the history of social anthropology.[2]

The leader of the expedition was the anthropologist and zoologist A. C. Haddon. Other members included W. McDougall, C. S. Myers, S. Ray, W. H. R. Rivers, C. G. Seligman, S. Ray, and A. Wilkin. Most important for this paper is the work of the Cambridge anthropologist, neurologist, psychiatrist and psychologist William Halse Rivers Rivers (1864-1922).[3] He investigated the perception and classification of colours on three islands, Mabuaig, Mer (= Murray Island), and Kiwai.

The central context for Rivers' investigations were earlier claims by English and German classical philologists according to which the human colour perception "as we know it" is of recent origin. For instance, William Ewart Gladstone (1809-1898), the four-time British prime-minister and Homer scholar, argued in 1858 – on the basis of philological evidence – that Homer and his audience were unable to distinguish colours, that is, that Homer and his audience was able only to distinguish between differences in brightness. The German philosopher and philologist Lazarus Geiger (1829-1870) later offered an evolutionary scheme according to which colour perception developed through distinct stages. Evolutionary development adds more and more colours in the order of the spectrum. The last colour that Westerners have learnt to discriminate has been blue.[4]

The ideas of Gladstone and Geiger were controversial by the time Rivers undertook his investigations. Rivers wanted to test these ideas experimentally. In setting out to do so, he assumed that the islanders of the Torres Strait were at a lower evolutionary stage than himself and his Cambridge experimental subjects. Ultimately Rivers thought that his work confirmed Geiger's and Gladstone's claims:

> "One of the chief interests of the work described in this report is that it shows that defect in nomenclature for a colour may be associated with defective sensibility for that colour and so far lends some support to the views of Gladstone and Geiger." (Rivers 1901a: 49)

---

2 Cf. Herle and Rouse 1996, Kuklick 1994, Saunders 2000, Slobodin 1978, and Stocking 1995: 98-114.
3 Cf. Slobodin 1978.
4 Cf. Kuklick 1994.

In order to get a sense of Rivers' "evidence", we need to take a closer look at his experiments and measurements. A first central tool for studying the colour discrimination of the islanders was "Holmgren's Wools", a device developed in the 1870s to test for colour blindness.[5] A large number of skeins of wools of different colour shades were poured in front of the experimental subject, and the subject was asked to sort them into seven piles. Each pile had one specific colour shade as a starting point or anchor: red, green, pink, pale ("Holmgren's") green, yellow, blue, and violet. Rivers reported the following results. First, "there was a natural tendency to put together all the wools to which the same name was given [...]" (Rivers 1901a: 49). Second, while his subjects did not match red to green or yellow to blue, "confusion between green and blue was very common and also between blue and violet" (Rivers 1901a: 51). Put differently, "the pale green wool [...] was matched correctly by the majority, but in a large number of cases it was matched with a number of bluish or violet wools [...]" (Rivers 1901a: 49).

A second experiment used "Lovibond's tintometer".[6] It consists of a tube with slots for inserting small pieces of coloured glass and an eyepiece at one end. In Rivers' set-up, the tintometer was directed at a white surface. Central to the experiment were "three series of coloured glasses, red, yellow and blue, very delicately graded so that each forms a series by means of which one passes from a colour so faint as to be indistinguishable from colourless glass up to a glass of a high degree of saturation." (Rivers 1901a: 71) The experimental subject was looking through the eyepiece while Rivers inserted different glasses along the described series. The subject had to tell Rivers as soon as she or he was able to identify a colour. The idea was to find the threshold at which colours were recognised correctly. Comparing the islanders' performance with that of English experimental subjects back in Cambridge, Rivers concluded that "the Murray Islander is relatively rather more sensitive to red than the Englishman, and distinctly less sensitive to blue." (Rivers 1901a: 73)

A third line of inquiry concerned colour nomenclature directly. Rivers asked the islanders to name the colours of standard sets of coloured papers, of various objects of the environment, of the colours in the tintometer, and of Holmgren's wools. The islanders on Murray Island gave one word for red: "mamamamam", but several words for blue: "bulu-bulu", "golegole", "suserisuseri", "gausgaus", "giazgiaz", "lulam gimgam", "akosakos", "soskepusoskep" ("colour adjectives in Murray Island are formed by reduplication from the names of various natural objects." (Rivers 1901a: 56). The crucial observation for Rivers was that "there was great definiteness and unanimity in the nomenclature for red [...] and very great indefiniteness for blue [...]" (Rivers 1901a: 54-5). Rivers also suggested an evolutionary perspective on the differences in colour nomenclature on the three islands: "As regards blue, the three languages [...represent] three stages in the evolution of a nomenclature for this colour." (Rivers

---

5 Cf. Collins and Drever 1925: 61-74 and Whipple 1924: 187-9.
6 Cf. Lovibond 1915.

1901a: 66) The inhabitants on Kiwai were thought to be at the lowest stage since they had no word for blue at all. The Murray Islanders were more advanced since they had adopted the term blue, or "bulu-bulu", from their English-speaking visitors. And the people of Mubaig were at a still higher level of development insofar as their word "maludgamulnga" "is used definitely for blue, but is also used for green". Alas, even the Mubaig Islanders still sometimes "confused" black and blue (Rivers 1901a: 66).

To sum up, Rivers was convinced that "the Papuan is characterized by a certain degree of insensitiveness to blue [...]" and that even "intelligent natives [...] regard it as perfectly natural to apply the same name to the brilliant blue of sky and sea which they give to the deepest black." (Rivers 1901a: 94) Rivers offered various explanations for why this might be so. A first explanation was physiological: the yellow pigmentation of the macula is greater in black-skinned than in white-skinned people, leading to a greater absorption of green and blue rays in the eyes of the former (Rivers 1901a: 79-80). A second explanation was that there are few blue pigments in the Torres Straits: "the nearest approach to a blue pigment was a bluish-grey clay" (Rivers 1901a: 96). Finally, in the third explanation the evolutionary scheme is again strong: "Another factor [...is] the absence of aesthetic interest in nature [...]. The blue of the sky, the [...] blue of the sea [...] do not appear to interest the savage." (Rivers 1901a: 96; cf. 64)

Rivers later sought to extend his investigations from the Torres Strait to other areas. In 1901 he published a paper on "The Colour Vision of the Natives of Upper Egypt" and in 1905 a study entitled "Observations on the Senses of the Todas", a tribe in Southern India. In both cases he claimed to have identified insensitiveness to blue.

Rivers' study was widely discussed in English-speaking anthropology and psychology; here are but a few examples of the subsequent debate. To begin with, other Cambridge psychologists also worked on colour, often influenced by Rivers' work. For example, C. S. Myers – another veteran of the Torres Strait expedition – published a paper on "the development of the colour sense" in 1908. In this study, Myers related the colour sensitivity of so-called "primitive people" to that of children. The lesser sensitivity for blue was explained in terms of "attraction": "Both primitive peoples and infants are attracted most by red and next by yellow [...]" (Myers 1908: 361-2). At the same time Myers rejected arguments like Gladstone's and (in part) Rivers' that drew conclusions about colours sensibility from linguistic data (Myers 1908: 361-2).

Robert S. Woodworth (1869-1962), the influential American psychologist wrote on "The Puzzle of Color Vocabularies" in 1910. Woodworth stressed the importance of *salience* for the development of different colour vocabularies:

> "Green and blue, in nature, are predominantly background colours, while red and yellow are usually the colours of small objects [...] recognized most readily by their colour. [...] Red and yellow are the usual colours of such important objects as ripe fruits, domestic animals, wild animals [...], blood and flesh." (Woodworth 1910: 333)

Woodworth also hypothesized that the availability of pigments had a significant impact on vocabularies of colour: "With the introduction of green and blue paints and dyes [...] names for green and blue have become stereotyped in European languages." (Woodworth 1910: 334) – Interestingly enough, Woodworth felt that these kinds of considerations made evolutionary explanations superfluous.

The most detailed discussion of Rivers' work in the Torres Strait was Edward Bradford Titchener's (1867-1927) paper from 1916: "On Ethnological Tests of Sensation and Perception with Special Reference to Test of Colour Vision and Tactile Discrimination Described in the Reports of the Cambridge Anthropological Expedition to Torres Straits". Titchener challenged Rivers on almost every point. For instance, Titchener suspected that the different colour discrimination thresholds of Islanders and Englishmen was an artefact of the different lighting condition in dark huts on the one hand, and a well-lit psychological laboratory in Britain. He also interestingly objected to Rivers' translations of native words and expressions. Rivers had translated "gole-gole" as "black" on the grounds that "gole" means cuttlefish, and that therefore "gole-gole" ought to stand for the colour of the cuttlefish's ink. Titchener did not find this translation compelling:

> "[...] the word *gole* means, not cuttle ink, but cuttlefish; and it is characteristic of these animals that they change colour, chameleonwise, to suit the colour of their surroundings. May it not be that the thought in the native's mind, when he uses the word *gole*, is "can't find him", "can't see him"? [...] And if this is the case, is it not natural that the adjective *golegole* should be applied to any large expanse within which no discriminable features can be made out? The dark of night, the skin of the body, the expanse of sea and sky [...]" (Titchener 1916: 224-5)

Of course, if "golegole" does not mean (primarily) black but "not having discriminable features" then the Torres Islanders are not "confused" in applying it to both black and blue surfaces.

Finally, Arthur Maurice Hocart's (1883-1939) "The 'Psychological Interpretation of Language'" from 1912 emphasised most forcefully the need to make sense of colour vocabularies against the background of cultures and their customs. Only against this background, Hocart insisted, could one meaningfully evaluate the "analytic strength" of a given nomenclature. For instance, in Central Asia horses are classified on the basis of their colour, and there is no general term for horse as such. For Hocart this fact alone tells us nothing about the analytic skills and general intelligence of the people living there. Indeed, Hocart claimed that their practice was not far from that of horsemen in his own English culture: "I think I am right in saying that a horsey person never speaks of a stallion or a mare as a horse." (Hocart 1912: 271; cf. Scuba 2002.)

As I mentioned in my introduction, Rivers' work is still important even today, at least as an influence on the important research project of Brent Berlin (1936-) and Paul Kay (1934-). As the two men write in their classic *Basic Color Terms*: "Rivers' work was the last attempt to discuss the evolution of colour nomenclature until the present study nearly seventy years later." (Berlin & Kay 1969/1999: 149; cf. Saunders 2000.)

basic colour terms are best understood *via negationis*. The following colour terms are not basic: "(a) crimson, (b) scarlet, (c) blond, (d) blue-coloured, (e) bluish, (f) lemon-coloured, (g) salmon-coloured, (h) the colour of rust on my aunt's old Chevrolet" (Berlin & Kay 1969/1999: 5). Berlin and Kay lay down a number of criteria to exclude all of (a) to (h). Such criteria include, for instance, that basic colour terms should be "monolexemic", "psychologically salient", or not specific to a narrow domain (Berlin & Kay 1969/1999: 6).

Berlin and Kay claim to have found that

> "[…] although different languages encode in their vocabularies different numbers of basic colour categories, a total universal inventory of exactly eleven basic colour categories exists from which the eleven or fewer basic colour terms of any given language are always drawn." (Berlin & Kay 1969/1999: 2)

These eleven colours are: white, black, red, green, yellow, blue, brown, purple, pink, orange, grey. Moreover, however many basic colours – two to eleven – a given language encodes, there are "strict limitations on which categories it may encode". The idea is best captured in the following well-known picture:

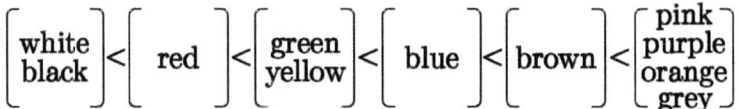

Basic colour terms; cf. Berlin & Kay 1969/1999: 4.

What to Rivers were Holmgren's Wools, Lovibond's Tintometer or standardized coloured papers, to Berlin and Key is the Munsell Colour Chart, a standardized way of ordering and numbering colours and their shades. Different colours and their shades are captured in a coordinate system, with hue on the horizontal axis, and chroma on the vertical. The data collection was then carried out as follows:

> "First, the basic colour words of the language in question were elicited from the informant, using as little as possible of any other language. Secondly, each subject was instructed to map both the focal point and the outer boundary of each of his basic colour terms on the array of standard colour stimuli [i.e. the Munsell Colour Chart]." (Berlin & Kay 1969/1999: 5)

3 Enter Wittgenstein

In this section I shall try to relate one strand of comments in Wittgenstein to the above work in the anthropology and psychology of colour. My first task is to make plausible that Wittgenstein knew about the work on colour by Rivers, Myers and others – even though he never mentions it explicitly in his writings.

Between 1911 and 1913 Wittgenstein did experimental-psychological work in Cambridge on the psychology of music, a field important to C.S. Myers (by then the head of the experimental psychology unit). Myers' project included a cross-cultural perspective. As we saw above, Myers also had interests in the psychology and anthropology of colour. Rivers was in Cambridge, too, but in anthropology. He was a fellow in St. John's College, and thus only a stone's throw from Trinity College where Wittgenstein was a student. Rivers and Myers were friends, and it is likely that Rivers visited the psychology unit at least occasionally. Wittgenstein knew Myers personally, and debated the relationship between psychology and logic with him. In July 1912 Wittgenstein read (as he wrote to Bertrand Russell) a "most absurd paper on rhythms" to the British Psychological Society when it met at Cambridge. He also demonstrated "an apparatus for psychological investigation of rhythm" at the opening ceremony of the new psychological lab in May 1913 (McGuinness 1988: 125-128, Pinsent 1990: 27). Add to this Wittgenstein's life-long and strong interest in other areas of anthropology – e.g. the work of Trinity fellow James Georg Frazer – and it seems hard to imagine that Wittgenstein did not learn a fair amount about the Torres Strait work already in the early 1910s.

Still stronger evidence emerges from even a superficial investigation into colour-related passages in Wittgenstein's oeuvre from the 1930s onwards. I begin with Wittgenstein's socio-cultural accounts of unexpected colour similarity judgements. He discusses a number of hypothetical cases in which members of other cultures make (to us) surprising similarity judgements concerning colour-samples. Wittgenstein offers rationales for the responses in terms of social or cultural factors – reminiscent of Hocart and Woodworth. And he does not invoke explanations in terms of colour-blindness as an evolutionary lag.

(A) Socio-cultural accounts of similarity judgements

Example I: (To us) red objects are judged by members of another culture to have something in common with, or be similar to, (to us) green objects. (Think of someone matching the green skein with a red skein in the Holmgren test.)

> "Red and green the same. [...] But don't the people see the difference?! Of course they do. But they have a word, say, "leaf-colour" [...it it] means red *or* green; and [...] two modifiers 'sharp' and 'blunt' [...] which separate red from green. [...] would these people be colour blind? Well, if we teach them our language they turn out to be normal." (LWPP I, 220)

Wittgenstein thus describes a possibility: the tribesmen apply the same term to what we call "red" and "green"; but they are able to learn our terms. Their use of the modifiers "sharp" and "blunt" in combination with their word for "red or green" shows that they are able to discriminate what we call "red" and "green". Note that Wittgenstein seeks to explain the tribesmen's vocabulary not in terms of an alleged cognitive deficit, or evolutionary lag, but in terms of their needs and aims: "It's just that the difference between red and green isn't as important to them as it is to us." (LWPP I, 221) Wittgenstein also makes the further suggestion that such classification might also plausibly find an expression in an artistic convention: "A type of painting, in which the illuminated side of figures is always painted green, the shadows always red" (LWPP I, 223). So much for Rivers' claim that not having an equivalent of our term "blue" must be an expression of a lack of aesthetic appreciation of nature. For Wittgenstein the key explanatory resource instead is salience and practical interests (cf. RPP I, 47 and 626; also WLPP, p. 115-238, in particular p. 121).

Example II: (To us) yellow objects are judged by members of another culture to be similar to (to us) blue objects; and (to us) green objects are judged by members of another culture to be similar to (to us) red objects.

> "Imagine a use of language (a culture) in which there was a common name for green and red on the one hand and yellow and blue on the other. Suppose, e.g., that there were two castes, one the patrician caste, wearing blue, red and green garments, the other, the plebeian, wearing blue and yellow. [...] Asked what a red patch and a green patch have in common, a man of our tribe would not hesitate to say that there were both patrician." (BB, p. 134)

If the members of this culture did use the same terms in these cases, Rivers, Gladstone and Geiger would presumably judge them to be on a lower evolutionary stage than Westerners. This is not Wittgenstein's position. He offers a social-cultural explanation that treats the other culture in a neutral, non-evaluative way. The members of the other culture act as we could imagine ourselves acting.

Example III: (To us) light blue objects are judged by members of another culture not to be similar to, and not to have something in common with, dark blue objects:

> "We could also easily imagine a language (and that means again a culture) in which there existed no common expression for light blue and dark blue, in which the former, say, was called 'Cambridge', the latter 'Oxford'. If you ask a man of this tribe what Cambridge and Oxford have in common, he'd be inclined to say 'Nothing'." (BB, p. 134-135)

Wittgenstein is not just picking these colours randomly: light greenish blue is called "Cambridge blue" and darkish blue "Oxford blue". Every-one in England is aware of this contrast (not least because of the annual Boat Race down the Thames). Of course Wittgenstein ironically pushes the example further than its real-world starting-point: he pushes it counterfactually to the point where Rivers would have said that the English are on a lower evolutionary scale than other Caucasians since at least in some contexts they do not emphasise the common features of light blue and dark blue.

(B) Differences in seeing

Unexpected similarity judgements to one side, Wittgenstein is also interested in cases where there is a temptation to describe differences in colour judgements between us and another culture as "differences in seeing". Such cases raise issues about symmetry, translatability and shared concepts. Let me explain.

(i) Consider a tribe with a complex mathematical system for identifying colours. We might call it the "tribe of the Lovibonds" to mark the similarity of Wittgenstein's thought experiment with the scaling of colours in Lovibond's tintometer:

> "Let us imagine men who express a colour intermediate between red and yellow, say by means of a fraction in a kind of binary notation like this: R, LLRL and the like, where we have (say) yellow on the right, and red on the left. [...] They would be related to us roughly as people with absolute pitch are to those who lack it." (Z, 368)

The first interesting feature here is the last sentence: people with absolute pitch can do something most of us are unable to do. By analogy, Wittgenstein seems to be suggesting that we should be careful about regarding our ordinary colour vision as the ultimate standard. There is at least the possibility that some people (individuals or other cultures) have colour vision that is superior to ours in more than one dimension. Or perhaps the point is this: we do not regard our musical ability deficient because we do not have absolute pitch. But then there is no good reason either to regard so-called "colour blindness" as a defect.

Wittgenstein is interested also in another aspect of this scenario: "We cannot imagine what it would be like to associate numerals with colours" (WLPP, p. 133). Or: "[...] a whole tribe [...] they give a set of numbers to shades of colours [...] One wants to say that one cannot imagine their experience" (WLPP, p. 253). Wittgenstein seems willing to allow for the possibility that there may be forms of colour experience that we can only characterise abstractly, but that we are unable to experience. The comparison with absolute pitch shows that there is no mystery here. After all, we "ordinary" people are not able to experience what it would be like to have absolute pitch either.

(ii) Another important thought experiment about "differences in seeing" concerns a tribe with "different colour poles" (WLPP, p. 19). For some purposes we use what Wittgenstein calls "colour poles"; terms around which we organise our "colour geometry". For us such colour poles are red, blue, yellow (and perhaps also green). We use these poles to define other colours. We say, for example, that purple is a bluish red. Assume now that there were a different culture with different colour poles, say purple, orange, blue-green, and yellow-green. And members of that other culture call red a purplish orange. Wittgenstein is doubtful whether such a language would be translatable into ours: "this tribe and we couldn't learn one another's language." And he likens this difference in colour poles to the difference between a colour-blind and a normally seeing person (WLPP, p. 19; cf. WLPP p. 138 and p. 258-259 as well as RPP I, 622). This suggests once more that for Wittgenstein the colour vision of people we call "colour-blind" need not be regarded as deficient; it is just a different way of organising the space of colours. The point is explicitly made elsewhere: "We speak of 'colour-blindness' and call it a *defect*. But there could easily be several different abilities, none of which is clearly inferior to others." (RC III, 31)

(iii) The third example of a difference in seeing is an encounter between us and a colour-blind tribe:

> "There could very easily be a tribe of people who are all colour-blind and who nonetheless live well; but would they have developed our colour names, and how would their nomenclature correspond to ours? What would their natural language be like?? Do we know? Would they perhaps have three primary colouPs: blue, yellow and a third which takes the place of red and green? – What if we were to encounter such a tribe and wanted to learn their language? We would no doubt run into certain difficulties." (RC III, 128; cf. RC I, 9, 13, and 66; also RC III, 32 and 154)

Wittgenstein's point about difficulties in translating the colour vocabulary of a tribe with a different organization of colours in noteworthy not least since it can be read as a criticism of Rivers. After all, Rivers and other authors of his tradition paid little attention to the difficulties of translating the language of a colour-blind tribe. Or put differently, they overlooked the difficulties that arise as we go from *one colour-blind person within our society to a whole tribe of the colour-blind*.

Although Wittgenstein considers the possibility of very different colour experiences and colour terms, he also feels the pull of our "universalistic intuitions":

> "Can't we imagine people having a geometry of colours different from our normal one? [...] The difficulty is obviously this: isn't it precisely the geometry of colours that shows us what we're talking about, i.e. that we are talking about colours?" (RC III, 86)

This consideration puts limits to how much variation in colour perception and colour nomenclature there could be.

(C) Languages without colour concepts

A further pertinent theme in Wittgenstein's reflections on colour concerns languages that do not have a distinct or separate system of colour terms. This idea has sometimes been used as a criticism of Berlin and Kay, and without reference to Wittgenstein.[7] Wittgenstein's first example is of a hypothetical linguistic community that, because of the character of its particular natural environment, has no need for colour concepts at all:

---

[7] Cf. Kuschel & Monberg 1974, Lucy 1997.

> "Suppose I were to come to a country where the colour of things – as I would say – changed constantly [...] The inhabitants never see unchanging colours. [...] It might be that their language *lacked* words for colours. [...] We might explain it by saying that they had little or no use for certain language-games." (RPP I, 198)

More realistic is a second case in which information about colour is encoded together with information about another dimension, here colour and form.

> "And what about people who only had colour-shape concepts? Should I say of them that they do not *see* that a green leaf and a green table – when I show them these things – have the same colour or have something in common? What if it had never 'occurred to them' to compare differently shaped objects or the same colour with one another?
> Due to their particular background, this comparison was of no importance to them, or had importance only in very exceptional cases, so that no linguistic tool was developed." (RC III, 130)

This tribe thus has one set of colour terms for leaves and a different set of colour terms for chairs. As Hocart (1912) reports in his above-mentioned paper, such "lack of abstractness" was one of the critical comments often directed at "savage" colour terminologies. Hocart of course countered this criticism with his emphasis on the need to consider the needs and environment of the respective language community. Wittgenstein clearly sides with Hocart (cf. RC III, 155; also MS 137, p. 8a-9b).

(D) Morphological differences

A final observation by Wittgenstein worth mentioning here can also serve as a critical observation on the idea of basic colour terms. He alludes to *morphological differences* in colour systems, for instance the possibility that colour terms might not be adjectives but verbs. ("In German we say '*es blaut*', [...] there might be a language where all colour words were verbs." (WLPP, p. 25)) As empirical research has shown, this phenomenon does indeed occur. The Berlin-Kay type of analysis pays insufficient attention to such variations.[8]

(E) A Grammar of Colour

At this point it is inviting to ask what Wittgenstein's alternative to the paradigm of Rivers, Berlin and Kay might be. Here one should keep in mind that, although Wittgenstein does not aim for a psychology, physiology or anthropology of colour, his suggestions might still be of substantive and methodological interest especially to the last-

---

[8] Cf. Lucy 1997.

mentioned field. Wittgenstein's goal is a "grammar of colour". This is best taken in the sense of the famous remarks from the *Investigations*, to wit that "grammar tells us what kind of object anything is" (PI, §373). Applied to this context, the grammar of colour judgements enables us to see and reconstruct our explicit and implicit commitments and assumptions about colours. I shall give a brief overview over some of this grammar's central pillars.

(i) A grammar of colour is a study of the "logic" of our colour concepts. "Logic" or "grammar" here contrasts with precisely the mentioned empirical studies: "We do not want to establish a theory of colour (neither a physiological one nor a psychological one), but rather the logic of colour concepts. [...]" (RC I, 22) That is to say, the focus is on the language games in which colour terms figure, not on individual-psychological or physiological abilities that allow us to discriminate between colours.

(ii) A grammar of colour focuses on what types of colour judgements are meaningful in what contexts. To understand these types one has to study both their role in various colour-related practices, and their relation to other types of judgements, both other types of judgements that pertain to colour, and to types of judgements concerning other dimensions, such as texture:

> "Describe the game with colours. The naming of colours, the comparison of colours, the production of colours, the connexion between colour and light and illumination, the connexion of colour with the eye, of notes with the ear, and innumerable other things." (RPP I, 628)[9]

(iii) It follows from (ii) that colour ascriptions are highly context-dependent:

> "If I say a piece of paper is pure white, and if snow were placed next to it and it then appeared grey, in its normal surroundings I would still be right in calling it white and not grey. It could be that I use a more refined concept of white in, say, a laboratory [...]." (RC I, 5)
> "I see in a photograph (not a colour photograph) [...] a boy with [...] blond hair [...] I see the boy's hair as blond [...] despite the fact that everything is depicted in lighter and darker tones of the photographic paper." (RC I, 63)

(iv) This context-dependence can also be expressed by saying that there is "*indeterminateness in the concept of colour or again in that of sameness of colour*" (RC I, 17). The two levels are alluded to in the following two passages:

> "Can a transparent green glass have the same colour as a piece of opaque paper or not? [...]" (RC I, 18)
> "It is easy to see that not all colour concepts are logically of the same sort, e.g. the difference between the concepts 'colour of gold' or 'colour of silver' and 'yellow' or 'grey'." (RC I, 54)

(v) A further important function of the grammar of colour is to identify dimensions of colour that have often been overlooked. As is familiar to every reader of *Remarks on*

---

9 Cf. Schneider 1978.

Colour, Wittgenstein became especially fascinated with the distinction between transparent and opaque colours. (This theme is extensively discussed elsewhere in this volume.[10])

(vi) Finally, Wittgenstein thinks that grammatical sentences concerning colour (i.e. rules governing the use of colour terms) behave similarly to other types of grammatical sentences: for instance, similarly to mathematical theorems. This idea is referred to when Wittgenstein speaks of a "geometry of colours" (RC I, 66) or when he writes: "We have a colour system as we have a number system" (RPP II, 426; Z 357, cf. LFM, p. 233-234.) To begin with, neither mathematical theorems nor grammatical sentences concerning colour are true in a correspondence-theoretical sense:

> "We have a colour system as we have a number system. Do the systems reside on our nature or in the nature of things? How are we to put it? – Not in the nature of numbers or colours." (RPP II, 426)

While not true in a correspondence-theoretical sense, grammatical sentences about colour, and thus systems of colour terms, can be more or less useful, given the needs of the respective language community. These sentences and systems are thus naturally related to "general facts of nature" about us and our way of life:

> "Then is there something arbitrary about this system? Yes and no. It is akin both to what is arbitrary and to what is not arbitrary." (Z 358)
> "'There is no greenish red' […] What would go wrong if we denied these laws? […] It would come to building a system which would be decidedly impractical." (LFM, p. 235)

Moreover, one and the same sentence, say, "this is red" can sometimes express a grammatical rule, and sometimes an empirical proposition "*so that their meaning changes back and forth*" (RC I, 32). Conflating these two uses leads to confusions – both here and in the case of mathematics (RC I, 88).

4 Conclusions

Above I have suggested that one important strand in Wittgenstein's reflections on colour are naturally read as responses to Rivers and other early twentieth-century psychologists and anthropologists working on the discrimination, perception and classification of colours. I have also claimed that Wittgenstein's criticism of Rivers might also – mutatis mutandis – be read as applying to aspects of Berlin's and Kay's work. Lest my main points are lost in the details, I shall conclude by highlighting my

---

10 Cf. Horner 2000.

main claims, and hinting at some implications for a broader perspective on Wittgenstein and the philosophy of the social sciences.

Wittgenstein's main points against Rivers and Berlin and Kay are the following. First, we must not diagnose a conceptual, intellectual or physiological defect in another culture simply on the grounds that its members draw conceptual lines (regarding colour) differently from us. Second, it would be a mistake to draw conclusions about how advanced a culture is on the back of its colour taxonomy. Third, phenomena we ordinarily classify as instances of colour-blindness need not be regarded as deficiencies in all contexts. Sometimes it might be more naturally to simply speak of a different organization of colours. Fourth, it is arbitrary to treat our colour taxonomy or vision as the standard or framework of analysis for all others. Fifth, Rivers underestimates the difficulties that might arise when we try to translate the colour terms of another culture. Sixth, if there are languages with form-colour concepts or with a different morphology then Rivers' methodology is not be applicable in a straightforward way.

In the past, reflections on Wittgenstein's importance for the philosophy of the social sciences in general and anthropology in particular have often focused on his "Comments on Frazer". If this paper is at least roughly near the mark, then this focus can now be widened by including a good number of Wittgenstein's remarks on colour into the corpus of relevant texts. The following more general precepts can then perhaps be formulated. A study of taxonomies, nomenclatures or vocabularies should be neutral and symmetrical in that the analyst does not evaluate linguistic phenomena, and in that he or she seeks to explain all of these phenomena on the grounds of their usefulness to particular cultures. Moreover, such study should focus on how language is used in ordinary contexts; a heavily constrained stimulus (be it skeins of wool, coloured glass, or colour chips) do not give us a good access to the rules underlying our language games. Furthermore, an investigation into taxonomies should not assume from the start that (what seem to us to be) distinct dimensions must be encoded in different and distinct categories. And finally, a satisfactory investigation into taxonomies must give special heed to the folk theories, or folk certainties, that surround and embed the central terms.[11]

Bibliography

Berlin, B. & Kay, P.: *Basic Color Terms*. CSLI Publications, 1969, 1999.
Bouveresse, J.: Wittgenstein's Answer to "What is Colour?". In: Moyal-Sharrock, D. (Ed.), *The Third Wittgenstein*, p. 177–191, Ashgate, 2004.
Brenner, W.: Wittgenstein's Color Grammar. In: *Souther Journal of Philosophy*, 20, p. 289–298, 1982.

---

[11] Work on this project was in part funded by an ERC Advanced Grant.

Collins, M. & Drever, F.: *Colour Blindness*. Kegan, 1925.
Herle, A. & Rouse, S. (Ed.): *Cambridge and the Torres Strait*. Cambridge University Press, 1996.
Hocart, A. M.: The "Psychological Interpretation of Language". In: *British Journal of Psychology*, 5, p. 267–279, 1912.
Horner, E.: "There cannot be a transparent white": A Defence of Wittgenstein's Account of the Puzzle Propositions. In: *Philosophical Investigations*, 23, p. 218–241, 2000.
Kuklick, H.: The Colour Blue: From Research in the Torres Strait to an Ecology of Human Behavior. In: MacLeod, R. & Rehbock, P. F. (Ed.), *Darwin's Laboratory*, p. 339–367, University of Hawai'i Press, 1994.
Kuschel, R. & Monberg, T.: "We don't talk much about colour here": A Study of Colour Semantics on Bellona Island. In: *Man*, 9, p. 213–242, 1974.
Lee, A.: Wittgenstein's *Remarks on Colour*. In: *Philosophical Investigations*, 22, p. 215–239, 1999.
Lee, A.: Colour and Pictorial Representation. In: *British Journal of Aesthetics*, 45, p. 49–63, 2005.
Lovibond, J. W.: *Light and Colour Theories and their Relation to Light and Colour Standardization*. Spon, 1915.
Lucy, J. A.: The Linguistics of "Color". In: Hardin, C. L. & Maffi, L. (Ed.), *Color Categories in Thought and Language*, p. 320–346, Cambridge University Press, 1997.
McGinn, M.: Wittgenstein's *Remarks on Colour*. In: *Philosophy*, 66, p. 435–453, 1991.
McGuinness, B.: *Wittgenstein: A Life. Young Ludwig 1889-1921*. Duckworth, 1988.
Myers, C. S.: Some Observations on the Development of the Colour Sense. In: *British Journal of Psychology*, 2, p. 353–362, 1908.
Pinsent, D. H. (Ed.): *Reise mit Wittgenstein in den Norden: TagebuchauszÂ͵ge - Briefe. Herausgegeben von G. H. von Wright*. Folio, 1990.
Rivers, W. H. R.: Colour Vision. In: Rivers, W. H. R. (Ed.), *Reports of the Cambridge Anthropological Expedition to Torres Straits. Volume II: Physiology and Psychology*, p. 48–96, Cambridge University Press, 1901.
Rivers, W. H. R.: Observations on the Senses of the Todas. In: *British Journal of Psychology*, 1, p. 321–396, 1905.
Rove, M. W.: Goethe and Wittgenstein. In: *Philosophy*, 66, p. 283–303, 1991.
Saunders, B.: Revisiting Basic Color Terms. In: *The Journal of the Royal Anthropological Institute*, 6, p. 81–99, 2000.
Schaffer, S.: Opposition is True Friendship. In: *Interdisciplinary Science Reviews*, 35, p. 277–290, 2010.
Schneider, J.: Peacocks and Penguins: The Political Economy of European Cloth and Colors. In: *American Ethnologist*, 5, p. 413–447, 1978.
Scuba, L.: Hocart and the Royal Road to Anthropological Understanding. In: *Social Anthropology*, 10, p. 359–376, 2002.
Slobodin, R.: *W. H. R. Rivers*. Columbia University Press, 1978.
Stocking, G. W.: *After Tylor: British Social Anthropology 1888-1951*. University of Wisconsin Press, 1995.
Titchener, E. B.: On Ethnological Tests of Sensation and Perception with Special Reference to Test of Colour Vision and Tactile Discrimination Described in the Reports of the Cambridge Anthropological Expedition to Torres Straits. In: *Proceedings o fthe American Philosophical Society*, 55, p. 204–236, 1916.
Vendler, Z.: Goethe, Wittgenstein, and the Essence of Color. In: *Monist*, 78, p. 391–410, 1995.
Westphal, J.: *Colour: Some Philosophical Problems from Wittgenstein*. Blackwell, 1987.
Whipple, G. M.: *Manual of Mental and Physical Tests. Part I: Simpler Processes*. Kessinger, 1924.
Woodworth, R. S.: The Puzzle of Colour Vocabularies. In: *Psychological Bulletin*, 7, p. 325–334, 1910.

Barry Stroud
Concepts of Colour and Limits of Understanding

Wittgenstein raises many difficult, puzzling questions in these very late notebooks published as *Remarks on Colour*. "Why is it that something can be transparent green but not transparent white?" (RC I, 19). "Why is there no brown or grey light?" (RC III, 215). "If I look at pure red through glass and it looks grey, has the glass actually given the colour a grey content...or does it only *appear* so?" (RC III, 207). "What does it mean to say 'Brown is akin to yellow'?" (RC III, 47).

I will not take up questions like this directly. Instead, I will discuss a more general concern Wittgenstein raises about our concepts of colours and the way our possession of those concepts does or does not impose certain limitations on our understanding of other apparent possibilities. This is something Wittgenstein struggled with not only about colours but about almost every other aspect of our thought.

I start with something that seems to have nothing to do with colour. Wittgenstein says "Someone who has perfect pitch can learn a language-game that I cannot learn" (RC III, 292). The language-game in question is presumably saying correctly what note you hear without relying on any independent standard. Wittgenstein cannot learn that language-game, he implies, because he does not have perfect pitch. Because of that limitation, which he shares with probably most other human beings, he cannot learn or participate in that practice. But although he does not have what it takes to engage in the practice, he presumably knows what the practice is. He knows and understands and can describe what those in that practice can do.

In some remarks collected in *Zettel*, probably from an earlier time than this material on colours, Wittgenstein draws a connection between perfect pitch and concepts of colours. He invites us to imagine people who speak of colours intermediate between red and yellow by means of fractions in a kind of binary notation representing different proportions of the colours at each end of the range from red to yellow. He presents this as a fully developed part of the life of these people. He says:

> These people learn how to describe shades of colour this way in the kindergarten, how to use such descriptions in picking colours out, in mixing them, etc.. They would be related to us roughly as people with absolute pitch are to those who lack it. *They can do* what we cannot. (Z 368)

We cannot do what those with absolute pitch can do, but we at least understand what they do. Do we understand what these people who speak of colours in this different way are doing or saying when they speak as they do?

Having described these people in this way, Wittgenstein says:

here one would like to say: "But then is it imaginable? Of course, the *behaviour* is! But is the inner process, the experience of colour?" And it is difficult to see what to say in answer to such a question. Could people without absolute pitch have guessed at the existence of people with absolute pitch? (Z 369)

What makes it difficult – if it is – to say whether or not this culture is imaginable does not seem to be whether we could have *guessed* that there are people who speak of colours in this different way. Even if we could not have guessed that there are some people with perfect pitch, we now know that some people have it. We understand very well what people with that gift can do; we even have a test for it. We have concepts of the musical scale in terms of which we understand and can recognize the achievement of those people.

The question about people described as speaking of the colours between red and yellow in these different ways is presumably whether we are in a position to understand what they do. Wittgenstein says what he calls their "behaviour" is imaginable, and so intelligible. His question is whether what he calls "the inner process, the experience of colour" of these people is imaginable. But whatever makes that question difficult, if it is, is surely not that we cannot gain direct access to the distinctive character of these people's "inner processes" or "experiences". The difficulty seems to lie rather in saying *what* experiences these people have when they see and speak in their special way of a certain colour. Wittgenstein says of these people that they can do what we cannot. But *what* can they do that we cannot? And can we understand what they do even if we cannot do it?

We recognize a range of colours intermediate between red and yellow – from a strongly reddish yellow to a strongly yellowish red. Those people Wittgenstein describes presumably do that too. Is it that they have many different *single words* for different points along that spectrum, while we have very few, or mostly only compound words? Is the "behaviour" we are to imagine them engaging in a matter of applying single words to a great many different colours in the range between red and yellow? If so, there is a question whether they really have different *concepts* of colour from ours. Of course, we did not learn as many different colour-words in the kindergarten and in subsequent training as they did. But their language-game, as described, seems to be a language-game we could learn. We seem to have the resources and the capacities already. At least it seems that we would not be excluded from their practice as most of us are hopelessly excluded from the language-game with perfect pitch. We can presumably make all the discriminations those imagined speakers make. So we would not be excluded from their practice in the way perceivers with various kinds of colour-blindness are excluded from full participation in the language-game of those with more or less normal vision. The colour-blind cannot make all the discriminations the rest of us can make.

Whatever can or cannot be understood about concepts of colours intermediate between red and yellow, Wittgenstein thinks things are radically different with something else we might think we can conceive of. He says:

> it is obvious at a glance that we aren't willing to acknowledge anything as a colour intermediate between red and green (Z 359).

It is not that red and green cannot be mixed; if you put them together you get *something*. The point is rather, as Wittgenstein puts it:

> we are not able to recognize straight off a colour that has come about by mixing red and green as one that can be produced in that way. (Z 365)

Does this mean we actually have no concept of a colour reddish green? At the very least, Wittgenstein says, we cannot see something as being that colour.

If that is so, is that to be understood as a limitation on our part? Is it like the limitation most of us are under of not having perfect pitch? If it is a limitation, it would be a limitation *everyone* is under, if what Wittgenstein says is right. But is it a limitation at all? If someone were to report that certain people are acquainted with reddish green, Wittgenstein imagines one might reply "But there *is* no such thing!". Wittgenstein is immediately struck by that reply: "What an extraordinary sentence", he says, "There *is* no such thing! (How do you know?)" (Z 362).

There would be nothing extraordinary in replying to a report that certain people are acquainted with round squares, for instance, by saying that there *are* no such things, and we know why. In finding the corresponding denial of such a colour as reddish green "extraordinary" I think Wittgenstein is struck by the oddness of denying that there is such a thing as a certain colour. What is it for no such colour to exist?

Even if we can make some sense of a language-game with concepts of colours intermediate between red and yellow, Wittgenstein thinks it is completely different with reddish green:

> can I describe the practice of people who have a concept, e.g. 'reddish green', that we don't possess? – In any case, I certainly can't *teach* this practice to anyone. (RC III, 122)
> Can I then only say: "These people call *this* (brown, for example) reddish green"? Wouldn't that just be another word for something I have a word for? If they really have a different concept than I do, this must be shown by the fact that I can't quite figure out their use of words. (RC III, 123)

This remark is immediately followed by the kind self-reproach found elsewhere in Wittgenstein's writings.

> But I have kept on saying that it's conceivable for our concepts to be different than they are. Was that all nonsense? (RC III, 124)

It looks as if it would be nonsense to speak of the conceivability of concepts different from ours if the only way it could be shown was by our being unable to make sense of the imagined speakers' uses of their words. That would not be to have conceived of a possibility of others possessing and using concepts recognizably different from ours. No determinate possibility at all would have been specified.

Nonetheless, does it not seem possible – conceivable – for our concepts to be, or at least to have been, different from what they are? Couldn't we who operate in all the ways we do have done certain things differently, and in ways we can now conceive of and understand? This is a more complicated question than it might seem. One way Wittgenstein puts the difficulty is:

> "Can't we imagine people having a different geometry of colour than we do?" – That, of course, means: Can't we imagine people who have colour concepts which are other than ours; and that in turn means: Can't we imagine that people do *not* have our colour concepts and that they *have* concepts which are related to ours in such a way that we would also want to call them "colour concepts"? (RC III, 154)

This last condition is of course essential. *We* have to be able to recognize the different concepts that we conceive of as colour concepts in order to acknowledge the imagined possibility of our *colour* concepts being different from what they are.

This means, as Wittgenstein says:

> We will, therefore, have to ask ourselves: What would it be like if people knew colours which our people with normal vision do not know? In general this question will not admit of an unambiguous answer. For it is by no means clear that we *must* say of this sort of abnormal people that they know other *colours*. There is, after all, no commonly accepted criterion for what is a colour, unless it is one of our colours. (RC III, 42)

We must be able to find something in what the people we conceive of can do that we can make sense of. And our resources for making sense of it, our resources for making sense of anything, lie in what *we* now know how to do and in our current ways of understanding things. All reflection or speculation about other conceivable possibilities faces a general kind of difficulty that Wittgenstein at one point describes this way:

> We say: "Let's imagine human beings who don't know *this* language-game". But this does not give us any clear idea of the life of these people, of where it deviates from ours. We don't yet know what we have to imagine; for the life of these people *is* [in other ways] supposed to correspond to ours ... and it first has to be determined what we would call a life that corresponds to ours under the new [imagined] circumstances.
>
> ... Questions immediately arise.... You have to make *further* decisions which you did not foresee in that first statement. (RC III, 296)

Further decisions or specifications are needed to fill out the description of the otherwise 'normal', 'familiar' human life the different ways of behaving we try to conceive of are to be imagined as fitting into. When the differences are small, or local, it is less clear that genuinely different *concepts* are involved. The colour-blind among us, for instance, cannot learn all the uses of colour words that the rest of us are masters of. But they understand a great many colour sentences; they just cannot use those sentences correctly in as many different ways as normal perceivers can (RC III, 278). It seems easier to make sense of others who are capable of *less* than what the rest of us know we can do, or even others who lack altogether certain concepts we know we have. That is to conceive of practices less rich or less complex than ours, not necessarily to conceive of concepts *different* from ours.

The difficulties of understanding the use of concepts different from ours arise more clearly in attempting to conceive of others who do not less but *more* than we do, so to speak, or something completely different. To conceive of others who have and use concepts that we do not even possess we have to describe more and more fully the practices in which those concepts get their sense. And as a growing number of further decisions have to be made to help specify more fully exactly what the possibility in question amounts to, we begin to lose our grip on what the people we are trying to conceive of are really doing, or how they understand it. Is what they can be understood to be doing really a conceptual alternative to what we understand ourselves to be capable of doing and saying in our familiar language-games?

When we press on for further specification, as we must, I think we can come to see that there is always much more at stake in the question of the intelligibility of concepts different from ours than appeared at the outset. Wittgenstein hinted in passing at part of the explanation of this in speaking of those with an elaborate vocabulary for colours intermediate between red and yellow as having leaned it in the kindergarten. He generalizes the point in a couple of remarks in *Zettel*.

> I want to say: an education quite different from ours might also be the foundation for quite different concepts. (Z 387)

> For here life would run on differently. – What interests us would not interest *them*. Here different concepts would no longer be unimaginable. In fact, this is the only way in which *essentially* different concepts are imaginable. (Z 388)

This says the only way we can really imagine concepts essentially different from our own is by imagining human beings with very different upbringing and education, whose life accordingly "runs on differently" from ours, and who show little or no interest in what interests us. If that is really what it takes, it is not clear in advance how far we can get in conceiving of the use of concepts different from ours. Can we really imagine or describe a form of human life in which we understand the concepts expressed in it to be radically different from ours? There seem to be no firm stopping-points in such imaginings; we carry on, having to add further qualifications or specifications. But still it seems – doesn't it? – that it is a contingent fact that we have the concepts and the conceptual resources that we do. So it seems hard to deny that, at least in very general terms, things could have been different from what they are.

The indeterminateness we encounter in trying to conceive of people with radically different concepts and ways of living comes in part from our pushing against the very conditions of our understanding anything. This is something Wittgenstein touches on in these notebooks but takes up more directly elsewhere and earlier. In *Remarks on Colour* he draws attention to the difference between learning to do something, or being master of a certain use or practice, and being able to describe or understand a description of that practice. He imagines certain "mental defectives who cannot be taught the concept 'tomorrow' or the concept 'I' " (RC III, 118). But then he asks:

> Now to whom can I communicate *what* this mental defective cannot learn? Just to whoever has learned it himself? (RC III, 119)

It looks as if the description of what such a person cannot do will be intelligible only to someone who has the very competence and understanding that the defective person is said to lack.

> How can I describe to someone how we use the word "tomorrow"? I can *teach* it to a child; but this does not mean I'm describing its use to him. (RC III, 122)

To understand the possibility of concepts other than ours we must be able to *describe* and understand the imagined practice in which those conepts are used. But "Understanding the description itself already presupposes that [the person who understands] has learned something", as Wittgenstein puts it (RC III, 121).

> I say to B, who cannot play chess: "A can't learn chess". B can understand that. – But now I say to someone who is absolutely unable to learn any game, so-and-so can't learn a game. What does he know of the nature of a game? (RC III, 282)

It seems that what is said about other people with different concepts will be intelligible only to someone who understands and is proficient in the use of the very concepts those people are said to make use of.

So when we ask, as Wittgenstein does, whether we can imagine people "having a geometry of colours different from our normal one", that means:

> can we describe it, can we immediately respond to the request to describe it, that is, do we know *unambiguously* what is being demanded of us?
> The difficulty is obviously this: isn't it precisely the geometry of colours that shows us what we are talking about, i.e. that we are talking about colours? (RC III, 86)

So to understand a description of what those with allegedly different concepts are doing and saying we must already be masters of language-games in which we can make sense of their doing what they do. We must be able to "domesticate" those people to some extent, so to speak – to find them intelligible in our own terms – on pain of not finding them intelligible at all. But to the extent to which we succeed in that effort, the people as described will not represent anything we can understand as radically different or 'other'.

I think there is an issue of more general significance behind all this that is worth drawing attention to. I think we come face to face here with the implications of an insight Wittgenstein expressed very early – in the 1930s – that remained fundamental to all his later thinking about understanding, meaning, and use. He wrote:

> The limit of language is shown by its being impossible to describe the fact that corresponds to (is the translation of) a sentence, without simply repeating the sentence. (CV, p. 13)

I take this to be directly relevant to the question of understanding the possibility of concepts different from ours. This remark says that it is impossible to say what a sentence says or means without making use of the very concepts expressed in that sentence. And it is impossible to understand what a sentence says without being competent in the use of those concepts. We could put it another way, as Wittgenstein sometimes does, by saying that it is only "from inside" understanding a language in which what the sentence says can be expressed that one can say and understand what the sentence says. So it is only "from inside" competence in and understanding of the structure of some conceptual practice or other that one can say what a particular sentence means or describe what people who utter and respond to that sentence are doing and saying.

If this is part of what lies behind the general difficulty of understanding concepts that are genuine alternatives to our own, and we succeed in that daunting task only to the extent to which we can conceive of human beings with upbringing, education, and basic interests and attitudes radically different from ours, it is perhaps not surprising that we cannot finally satisfy ourselves one way or the other about their conceivability. But if we can begin to understand why that is so, why do we continue to raise the apparently unpromising question of the possibility of concepts different from ours? What is at stake?

Is it perhaps that we think we can really understand our thinking and doing things in all the ways we do only if we can come to see how those ways of thinking and behaving are 'grounded' in or legitimately related to the world they enable us to think about?

Wittgenstein asks in *Zettel*:

> Do I want to say, then, that certain facts are favourable to the formation of certain concepts; or again unfavourable? And does experience teach us this? It is a fact of experience that human beings alter their concepts, exchange them for others when they learn new facts; when in this way what was formerly important to them becomes unimportant, and *vice versa*. (Z 352)

Or, as he puts it in *Remarks on Colour*:

> We could say people's concepts show what matters to them and what doesn't. But it is not as if that *explained* the particular concepts we have. It is only to rule out the view that we have the right concepts and other people the wrong ones. (RC III, 293)

This talk of "the right concepts" echoes the well-known remark at the end of what used to be called Part II of *Philosophical Investigations,* now called 'Philosophy of Psychology – A Fragment':

> If anyone believes that certain concepts are absolutely the correct ones, and that having different ones would mean not realizing something that we realize – then let him imagine certain very general facts of nature to be different from what we are used to, and the formation of concepts different from ours will become intelligible to him. (PI, p. 241e)[1]

This is Wittgenstein's target: the wish we apparently have to somehow vindicate our concepts or to show that they are "correct", the "right" or the "only" concepts to have in a world like this. The demand or hope is that our concepts can be matched up with, or be seen or shown to have their ground or basis in, identifiable features or aspects of the independent world the concepts are used to think and speak about. If that could be done, others with radically different concepts would be recognizable as missing something – not just something that we realize, but missing something that is really so in the way things are. Some such aspiration to understand the 'ground' or 'basis' of our concepts is what I think lies behind Wittgenstein's provocative, challenging remarks in *Zettel*.

> We have a colour system as we have a number system. Do the systems reside in *our* nature or in the nature of things? How are we to put it? – *Not* in the nature of numbers or colours. (Z 357)

> Then is there something arbitrary about the system? Yes and No. It is akin both to what is arbitrary and to what is non-arbitrary. (Z 358)

What does this 'answer' really say? That our "systems" are 'akin' to what is arbitrary, and 'akin' to what is non-arbitrary? But kinship, after all, is a very loose relation; it can extend great distances, and can even hold between incompatible extremes.

One can see why Wittgenstein says what he says here, but I think there is something unsatisfyingly asymmetrical in the way he leaves it. To the question whether the number system and the colour system "reside in *our* nature or in the nature of things?" he answers "*Not* in the nature of numbers or colours". That seems to mean that we cannot say that our system of numbers or colours is as it is because of the very nature of those things that the systems so successfully capture and express – numbers

---

[1] I discuss some implications of this remark and explore some of Wittgenstein's examples of different concepts or practices in my 'Wittgenstein and Logical Necessity', *Philosophical Review*, 1965 (reprinted in my *Meaning, Understanding, and Practice*, Oxford University Press, Oxford, 2000).

and colours. So we cannot say that different systems or ways of thinking of numbers or colours would miss something essential to those things we think about in our systems. It is not that the 'ground' of those systems of ours lies or "resides" in the nature of the things we think about in those ways.

That is all Wittgenstein says here, but it is an answer to only one half of what is a two-part question. If the systems do not "reside" in the "natures" of those things themselves, are we to take it that the systems of numbers and colours do "reside in *our* nature" instead? Wittgenstein says nothing about that apparent possibility one way or the other. He does not conclude from his negative answer to the first half of the question that the number and the colours systems do "reside in *our* nature". To draw that conclusion would presumably mean that the "geometry of colours", the structural relations that hold among the different colours we acknowledge and understand, could be different, or would have been different if *we* had been different in certain ways. Is that what our accepting the system's "residing in *our* nature" would amount to? If so, and if the system of numbers also "resides" not in the nature of numbers but in *our* nature, does that mean that what we now take to be truths of number would not have been true, or would have been different from what they are, if *our* nature had been different in certain ways?

To accept this alternative seems no better as an answer to the question than the appeal to the "nature" of numbers and colours that Wittgenstein rejects. Can we accept that if *we* had been different in certain ways there might have been different truths about numbers and colours, or perhaps no systems of numbers or colours at all? If the necessities that hold in those systems as we now have them are truly necessary, we must acknowledge that they would hold in *all* circumstances, and so would hold even if we and our nature had been very different from what they are. I think we cannot consistently acknowledge that what we regard as genuine necessities could find their source in *us*. Nor would their residing in the very "natures" of numbers and colours themselves help explain the necessity of the necessities we recognize either.[2]

The very question itself about the 'grounds' of our concepts and their associated necessities is what seems to lead to the troubles and dissatisfactions we encounter. In asking whether or to what extent we can conceive of the use of concepts of number or of colour different from ours we seem unable to find a satisfactory resolution either way, given what we already think and know about numbers and colours. It is the very thoughts we now have about numbers and colours that we wish to have some deeper or more illuminating understanding of. But what exactly *is* that question about the 'grounds' or 'basis' of the concepts we know we have? And what could ever put us in a position to answer such a question?

---

[2] I have tried to show why neither kind of answer can give a satisfactory account of several kinds of necessity or of value in my *Engagement and Metaphysical Dissatisfaction: Modality and Value*, Oxford University Press, New York, 2011.

Notes on the Contributors

ANDREW LUGG received a B.Sc (Eng) from University College London and an M.S.E and a Ph.D. (in philosophy) from the University of Michigan. He taught at the University of Ottawa from 1973 to 2002. His work in philosophy initially focused on questions about the nature of scientific inquiry, later on the history of twentieth-century analytic philosophy, in particular the writings of Ludwig Wittgenstein and W. V. Quine. Lugg is the author of Wittgenstein's Investigations 1-133 (2000/2004) and numerous articles in scholarly journals. In recent years he has been especially concerned with Wittgenstein's writings on colour.

JOACHIM SCHULTE studied philosophy, sociology, and musicology at the universities of Bonn, Cologne and Frankfurt. He did his graduate studies in philosophy at Oxford, where he took his D.Phil. He has taught at the Universities of Bologna, Graz, Bielefeld and Zürich. He is known for his translations of philosophical and historical as well as literary works from English and Italian into German. His publications include numerous articles and four books on Wittgenstein as well as critical editions of his main works. He has worked extensively on Ludwig Wittgenstein's Nachlass, and is a member of the board of editors of Wittgenstein's Nachlass at Trinity College, Cambridge. Joachim Schulte is co-editor and co-translator of the revised 4th edition of Wittgenstein's Philosophical Investigations. Currently, he is working chiefly on Wittgenstein's last writings.

RICHARD HEINRICH studied philosophy at the University of Frankfurt and the University of Vienna where he held a teaching position until his retirement in 2013. His research interests include but are not limited to aesthetics, the history of art, philosophy of mathematics, and the history of philosophy, in particular 17th and 18th century philosophy. His work on Ludwig Wittgenstein focuses mainly on issues relating to the Tractatus logico-philosophicus and the relation of Wittgenstein's early philosophical ideas to the writings of Immanuel Kant and Gottlob Frege. Currently he is working in the field of philosophy of literature.

GABRIELE M. MRAS is professor of philosophy at the Vienna University of Economics and Business (WU Wien) and teaches philosophy at the University of Vienna. She has held visiting professorships at numerous universities, among them the University of California, Berkeley and the University of Minnesota, Minneapolis. She has published books on early analytic philosophy, Frege and Russell and articles on Wittgenstein, in particular on his treatments of questions concerning indexicals, demonstratives, and the ascription of colour and number concepts. Currently she is working on a book on Russell and Bradley on the unity of judgement.

GARY KEMP received his PhD from the University of California at Santa Barbara. He has taught at the University of Waikato, the University of Glasgow, and the Vienna University for Economics and Business. He predominantly worked in the philosophy of logic and language, and on Frege, Russell, Quine, and Davidson, including a book on the last two. He has also published papers on aesthetics and philosophical themes in literature. More recently, he has published a paper comparing the later Wittgenstein's views with Quine's naturalism, and is currently working on a paper assessing Quine's attitude towards metaphysics.

FREDERIK A. GIERLINGER studied economics, philosophy, mathematics and pedagogy in Vienna. He has taught introductory courses in philosophy at the University of Vienna and was a Visiting Researcher at the University of California, Berkeley and Rutgers University in New Jersey. His areas of specialization are philosophy of language, metaphysics, and epistemology. He has published mainly on issues relating to the philosophical writings of late Ludwig Wittgenstein. Currently he is working on the status of moral judgments and questions concerning moral development.

HERBERT HRACHOVEC studied history, philosophy, and theology, as well as Germanic languages and literature in Vienna and Tübingen. He held fellowships and visiting appointments at the universities of Oxford, Münster, Harvard, Berlin, Essen, Weimar and retired from his position at the Department of Philosophy at the University of Vienna in 2012. His publications cover a wide range of topics with a focus on analytic philosophy, metaphysics, and aesthetics. With respect to Wittgenstein he has written on issues relating to his Nachlass and on a broad selection of themes in his philosophy, early and late. In recent years the focus of his work has been the philosophy of technology and issues relating to new media.

MARTIN KUSCH is professor for philosophy of science and epistemology at the University of Vienna. Prior to coming to Vienna in 2009, he held a chair in philosophy and sociology of science at the University of Cambridge. His main book publications are Language as Calculus vs. Language as Universal Medium (1989), Foucault's Strata and Fields (1991), Psychologism (1995), Psychological Knowledge (1998), The Shape of Actions (with H. M. Collins, 1999), Knowledge by Agreement (2002), and A Sceptical Guide to Meaning and Rules: Defending Kripke's Wittgenstein (2006). He is currently completing a book entitled Wittgenstein's Epistemological Investigations. In 2014 he was awarded an ERC Advanced Grant for a project entitled "The Emergence of Relativism: Historical, Philosophical and Sociological Perspectives".

Barry Stroud is Willis S. and Marion Slusser Professor of Philosophy at the University of California, Berkeley. He received his B.A. in philosophy from the University of Toronto and a Ph.D. from Harvard University. He has written extensively on philosophical skepticism, David Hume, and the later philosophy of Wittgenstein, and more recently on the metaphysical status of colours, causality, necessity, and values.

Index

colour ascription, 51, 53, 104
    predication, 45–47, 51, 53, 54
colour blindness, 2, 68, 69, 71, 74, 95, 99, 101, 102, 106, 110
colour model
    colour circle, 8, 40, 41, 52
    colour double cone, 8
    colour double pyramid, 8, 33, 34
    colour octahedron, 8–10, 13, 34, 37, 42, 45–49, 51, 53
    colour wheel, 18, 57
    perspicuous representation, 13, 17, 34, 46
colour vocabulary, 64, 106
    basic colour terms, 97, 98, 103
    nomenclature, 94, 95, 97, 102, 106
    taxonomy, 106

geometry of colour, 8, 18, 33, 41, 45, 46, 48, 53–55, 101, 102, 105, 112, 114, 117
    colour space, 33, 34, 37, 41, 42, 85
    geometrical relation, 45, 46, 48
    logic, 8, 13, 15, 16, 18, 26, 53, 57, 58, 65, 82, 84, 86, 104
    logical relation, 17, 36
    mathematics, 18, 27, 65, 82
    structural relation, 117
    structure, 33, 37, 42, 46, 48, 85, 86
grammatical rule, 8, 9, 48, 84, 105

impossible colour
    colour exclusion, 47, 83, 85
    complementary colour, 50, 85
    reddish-green, 4, 8, 10, 16, 33, 41, 46–48, 67, 68, 71–76, 79, 81–83, 85, 87–90, 111
    yellowish-blue, 22, 41, 47, 67, 68, 71–76

limits (of understanding)
    limits (of conceivability), 46, 112, 115
    limits (of language), 115
    limits (of possibility), 47
logical form, 81, 89
logical forml, 80

mixture, 37–42, 54
    blend, 6, 7, 9, 79
    mixing, 28, 109, 111
    order, 94, 98
    ordering, 11, 23, 98
    transition, 31, 41, 67, 73, 84, 85

nature of colour, 21, 22, 57–59, 90
    essence of colour, 58, 59, 61
negation, 26, 47, 50, 51, 83, 98
number, 65, 85, 105, 116, 117

perception of colour, 73, 85, 93, 94, 102, 105
    colour experience, 42, 58, 73, 101, 102, 110
phenomenology, 9, 39, 58, 64
    phenomenalist, 58, 59
physiology, 29, 73, 74, 103
    physiological, 28, 94, 96, 104, 106
primary colour, 16, 24, 42, 46, 49, 102
psychology, 34, 93, 96, 98, 99, 103
    Gestalt psychology, 23, 28, 29

sameness (of colours), 49, 58, 103, 104
    identity (of colours), 53
shade of colour, 6, 25, 27, 53, 95, 98, 101, 109
    colour and shadow, 54, 68, 69

transparency, 1, 9–18, 42
    behindness, 12, 13

cloudy, 6, 7, 13
colouredness, 6, 7
depth, 12, 13
opacity, 11–13
opaque, 6–13, 15, 104, 105

saturated, 54
saturation, 11, 95
translucent, 65
transparent, 6–18, 53, 65, 104, 105, 109

Names

Anscombe, G. E. M., 1–5, 25

Baker, G., 45, 46
Berlin, B., 97, 98, 102, 103, 105, 106
Bouveresse, J., 83
Brentano, F., 54, 93

Carnap, R., 64
Cavell, S., 74

Frazer, J. G., 99

Geiger, L., 94, 100
Gladstone, W. E., 94, 96, 100
Goethe, J. W. von, 3, 6, 7, 10, 12, 18, 93

Hacker, P., 45, 46
Haddon, A. C., 94
Hocart, A. M., 97, 99, 103
Hoefler, A., 33

Köhler, W., 29
Katz, D., 93
Kay, P., 97, 98, 102, 103, 105, 106
Kripke, S., 58

Lichtenberg, G. C., 93

Malcolm, N., 3, 4, 6, 7
McAlister, L., 26
McDougall, W., 94
McGinn, M., 9, 13
Myers, C. S., 94, 96, 98, 99

Nyman, H., 10

Paul, D., 3
Putnam, H., 58

Quine, W. V. O., 62

Ray, S., 94
Rhees, R., 1, 3, 4, 6, 7, 10
Rivers, W. H. R., 94–100, 102, 103, 105, 106
Rosefeldt, T., 75
Runge, P. O., 10, 11, 16, 33, 37–39, 42, 93
Russell, B., 99

Schättle, M., 26
Schaffer, S., 93
Schlick, M., 35, 36, 38, 47, 48
Schopenhauer, A., 93
Schulte, J., 54
Shakespeare, W., 4

Titchener, E. B., 97

von Wright, G. H., 1–3, 5, 6, 10

Waismann, F., 35, 36, 38, 47, 48
Watson, W.H, 8
Westphal, J., 33
Wilkin, A., 94
Woodworth, R. S., 96, 97, 99

www.ingramcontent.com/pod-product-compliance
Lightning Source LLC
Chambersburg PA
CBHW050913160426
43194CB00011B/2383